Introduction to Metalens Optics

Online at: https://doi.org/10.1088/978-0-7503-5784-5

IOP Series in Emerging Technologies in Optics and Photonics

Series Editor

R Barry Johnson, a Senior Research Professor at Alabama A&M University, has been involved for over 50 years in lens design, optical systems design, electro-optical systems engineering, and photonics. He has been a faculty member at three academic institutions engaged in optics education and research, has been employed by a number of companies, and has provided consulting services.

Dr Johnson is an IOP Fellow, an SPIE Fellow and Life Member, an OSA Fellow, and was the 1987 President of SPIE. He serves on the editorial board of *Infrared Physics & Technology* and *Advances in Optical Technologies*. Dr Johnson has been awarded many patents, has published numerous papers and several books and book chapters, and was awarded the 2012 OSA/SPIE Joseph W Goodman Book Writing Award for Lens Design Fundamentals (second edition). He is a perennial co-chair of the annual SPIE Current Developments in Lens Design and Optical Engineering Conference.

Foreword

Until the 1960s the field of optics was primarily concentrated in the classical areas of photography, cameras, binoculars, telescopes, spectrometers, colorimeters, radiometers, etc. In the late 1960s optics began to blossom with the advent of new types of infrared detector, liquid crystal display (LCDs), light emitting diode (LEDs), charge coupled device (CCDs), laser, holography, and fiber optics along with new optical materials, advances in optical and mechanical fabrication, new optical design programs, and many more technologies. With the development of the LED, LCD, CCD, and other electro-optical devices, the term 'photonics' came into vogue in the 1980s to describe the science of using light in the development of new technologies and the operation of a myriad of applications. Today optics and photonics are truly pervasive throughout society and new technologies are continuing to emerge. The objective of this series is to provide students, researchers, and those who enjoy self-education with a wide-ranging collection of books, each of which focuses on a topic relevant to the technologies and applications of optics and photonics. These books will provide knowledge to prepare the reader to be better able to participate in these exciting areas now and in the future. The title of this series is *Emerging Technologies in Optics and Photonics*, in which 'emerging' is taken to mean 'coming into existence', 'coming into maturity', and 'coming into prominence'. IOP Publishing and I hope that you will find this series of significant value to you and your career.

A full list of titles published in this series can be found here: https://iopscience.iop.org/bookListInfo/emerging-technologies-in-optics-and-photonics.

Introduction to Metalens Optics

Ivan Moreno

LUMAT, Universidad Autonoma de Zacatecas, Mexico

IOP Publishing, Bristol, UK

ISBN   978-0-7503-5784-5 (ebook)
ISBN   978-0-7503-5782-1 (print)
ISBN   978-0-7503-5785-2 (myPrint)
ISBN   978-0-7503-5783-8 (mobi)

DOI   10.1088/978-0-7503-5784-5

Version: 20240701

IOP ebooks

British Library Cataloguing-in-Publication Data: A catalogue record for this book is available from the British Library.

Published by IOP Publishing, wholly owned by The Institute of Physics, London

IOP Publishing, No.2 The Distillery, Glassfields, Avon Street, Bristol, BS2 0GR, UK

US Office: IOP Publishing, Inc., 190 North Independence Mall West, Suite 601, Philadelphia, PA 19106, USA

Cover image: "Metalens: The Light's Whisper" by Pablo Moreno.

To Sofia

Contents

Preface

To the best of my knowledge, this is the first published book dedicated solely to metalenses, perhaps because it is an emerging field. While a few book chapters have covered specific aspects of metalenses, a comprehensive text has been lacking. The primary objective of this book is to provide an introductory text about the emerging and fascinating field of metalenses. Through this book, readers will learn the basic optics of their operational mechanism, physical principles, simulation methods, and basic design methods. This book may serve as a reference for research scientists, optical engineers, and graduate and undergraduate students interested in flat optics or meta-optics, with a particular focus on metalenses.

This book is structured as follows: chapter 1 provides an overview of metalenses, spanning from their historical origins to the latest advancements in the field. Chapter 2 introduces the fundamental principles of metalens optics, offering readers an optical perspective on these innovative devices. In chapter 3, I explore software tools utilized for simulating metalenses and meta-atoms, providing readers with practical insights into computational techniques. Chapter 4 focuses on the principles of meta-atoms, shedding light on their underlying mechanisms and characteristics. Finally, chapter 5 discusses the principles of optical design for metalenses, offering valuable insights into the basic process of designing and optimizing these optical devices. Throughout each chapter, I aim to present foundational principles, key formulas, and essential concepts essential to understanding metalens optics comprehensively. It is my sincerest hope that readers will find this book both informative and thought-provoking as they explore the fascinating world of metalens optics.

Ivan Moreno
Zacatecas, Mexico

Acknowledgements

I extend my heartfelt gratitude to Dr Sofia Gamboa-Duarte for her unwavering support and encouragement during the writing of this book. Her motivation was instrumental in its completion. I also thank Professor R Barry Johnson for considering me for this endeavor and for his great idea to write a book about metalenses.

I am grateful to the members of my research group (Jose Carlos, Polet, Diana, David, Thaire, and Edgar) for providing me with invaluable opportunities to think deeply into the subjects of metasurfaces and metalenses through our numerous meetings at the Unidad Académica de Ciencia y Tecnología de la Luz y la Materia (LUMAT). Additionally, I express gratitude to Dr Hossein Alisafaee for valuable discussions on the subject of metalens design during a summer stay at Rose-Hulman Institute of Technology.

Lastly, I wish to express my deepest thanks to my parents (Maria de Jesus, and Vicente) for their unwavering support and the invaluable life lessons they have imparted to me.

Author biography

Ivan Moreno

Ivan Moreno is a specialist in optics and photonics. He received his BS degree in Physical Engineering from the Technological Institute of Monterrey (ITESM) and his PhD in Optical Sciences from the Optical Research Center (CIO), both in Mexico. He was the pioneer of LED lighting research in Mexico and is a researcher and educator at the LUMAT center at the Autonomous University of Zacatecas, Mexico. He has over 20 years of teaching experience in optics courses.

Dr Moreno was awarded the 2011 ICO-ICTP Gallieno Denardo Award. He has authored scientific articles in the areas of metalenses, metasurfaces, LED lighting, illumination, color, optical design, vision, thin-film filters, and interferometry. He is a member of the Mexican Academy of Sciences and a Senior Member of both the OPTICA and SPIE societies. He serves on the editorial board of the scientific journal Optics Express.

Abbreviations

2D	Two-dimensional
3D	Three-dimensional
CMOS	Complementary metal–oxide–semiconductor
FDTD	Finite difference time domain
FOV	Field of view
FWHM	Full width at half maximum
LED	Light-emitting diode
NA	Numerical Aperture
OPD	Optical path difference
OPL	Optical path length
PSF	Point spread function
RCWA	Rigorous coupled-wave analysis
TIR	Total internal reflection
Unit cell	The unit cell is the basic periodic unit that contains one nanoelement. While the nanoelement itself is not periodic, the unit cell, along with the substrate segment, is periodic.
vs	Versus

Symbols

λ	Wavelength (nm)
$\boldsymbol{E}$	Electric field (V m^{-1})
$\boldsymbol{B}$	Magnetic field (T)
Φ	Phase; also, phase shift
θ	Angle
k	Wave number (m^{-1})
D	Metalens diameter
R	Metalens radius
$D(r)$	Nanopillar diameter in function of radial distance
r	Radial distance from the metalens center
f	Focal distance
$f/\#$	f-Number
n	Refractive index
c	Speed of light in a vacuum
ε	Permittivity

IOP Publishing

Introduction to Metalens Optics

Ivan Moreno

Chapter 1

Introduction

Unlike traditional lenses, a metalens represents a revolutionary advancement in optics. It is a flat nanostructure comprising multiple meta-atoms intricately arranged into specific configurations known as metasurfaces (figure 1.1). Endowed with both focusing and imaging capabilities, metalenses hold immense potential for a wide array of applications, particularly in compact optical systems like mobile devices.

At their core, metalenses are a subset of metasurfaces tailored for focusing and imaging applications. They consist of meticulously arranged meta-atoms on a lattice, serving as the building blocks for shaping optical waves with exceptional spatial resolution. Early implementations of metalenses utilized ultrathin metallic meta-atoms; however, these designs were constrained by absorption losses and efficiency limitations. Recent advancements have shifted toward lossless dielectric meta-atoms, offering improved performance and practicality.

This chapter embarks on a captivating journey through the history of metalenses, tracing their origins to the present day, where cutting-edge research drives innovation in this dynamic field where nanotechnology and optics merge. Along the way, we explore the motivations behind the recent developments of metalenses and unveil the potential advantages they offer. Finally, we briefly describe the main optical characteristics that distinguish metalenses, laying the foundation for the deeper exploration that lies ahead in this book.

1.1 Historical background

The journey of metalens optics begins with the rich history of lens development, tracing back to antiquity. The term 'lens' finds its etymological roots in the Latin, which refers to the lentil seed, shaped akin to a double-convex lens [1]. The earliest documented use of lenses dates back to the ancient civilizations, with artifacts like the Nimrud lens, a rock crystal artifact from the 7th century BCE for the magnification and light manipulation. References in ancient Greek writings attest to the early recognition of lenses for magnification and correction of vision.

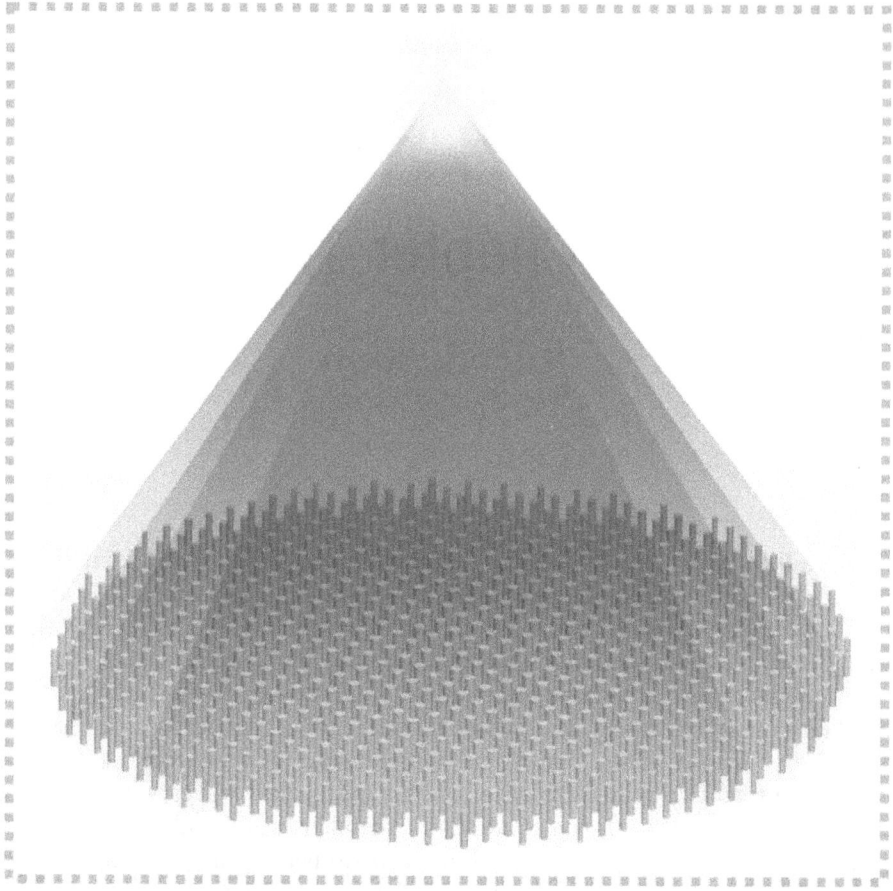

Figure 1.1. An optical metalens is a flat plate made of nano-optical elements with light focusing properties.

The medieval Islamic world played a pivotal role in advancing optical theories, with scholars like Ibn Sahl and Alhazen making significant contributions to our understanding of light refraction and lens behavior. The Renaissance and Early Modern periods witnessed remarkable progress, with luminaries like Johannes Kepler, René Descartes, and Isaac Newton laying the foundations of modern optics. From the invention of the telescope and microscope to the development of modern optics, the journey of lens technology has been marked by continuous innovation and discovery, paving the way for the emergence of metasurfaces in the 21th century. This historical backdrop sets the stage for exploring the transformative potential of the metalens in the modern era of optics.

The historical background of metalenses is a proof of the evolution of optical technology, tracing its roots from microwave diffractive lenses in the Second World War to cutting-edge developments in visible wavelengths at the present [2, 3]. Initial microwave diffractive lenses obtained high effective index by doping polystyrene foam sheets with subwavelength metallic insets (antennas), offering weight reduction and broader frequency operation. However, challenges persisted for decades in extending these principles to shorter wavelengths like visible light.

One could say that the first metalens in the visible was a plasmonic flat lens [4–7]. An early work demonstrated the focusing of visible light using quasiperiodic arrays of nanoholes in a metal plate [4]. In this work, focusing was achieved not via phase shifts, but by amplitude modulation and quasicrystal spatial distribution of nanoholes. However, a metalens is an ultrathin and flat plate that introduces a phase shift profile capable of focusing light. For such a task, a metalens requires a 2π phase coverage with ideally perfect transmittance for all phase shifters. In this context, a theoretical demonstration of focusing visible light by phase shifts using periodic arrays of nanoholes in a metal plate was reported in [5]. In this work, phase modulation was achieved by varying the depth (propagation length) of plasmonic slits, resulting in a topographic profile; however, this ultrathin lens was not yet flat. One might argue that the first metalens in visible, was a 2D metalens consisting of an array of slits with locally varying width, which was demonstrated by simulations [6]. In that theoretical research the phase profile was produced by adjusting the width of slits, i.e. the narrower the slit, the greater the phase shift. This early metalens concept was experimentally demonstrated through nanoscale slit arrays in a metallic film under red light illumination [7]. That early metalens structure consisted of a gold film with micron-size arrays of closely spaced, nanoscale slits of varying widths.

Breakthroughs in the infrared and visible domains, such as the use of graded subwavelength gratings and transparent high-index materials like TiO_2, paved the way for efficient metalenses in visible wavelengths. These advancements led to the fabrication of metalenses with unprecedented performance, surpassing traditional diffractive components, offering new avenues for controlling phase and achieving high focusing efficiency. The historical journey of the metalens highlights the continuous innovation and paradigm shifts driving the field of optics towards revolutionary applications in the realm of nanotechnology.

The history of the metalens is a tale of relentless innovation, as described by Professor Federico Capasso, a pioneering figure in this amazing optics field [8]. The journey towards efficient optical metalenses began with an optics need in sensing drones, where bulky lenses limited drone-mounted sensors. Driven by the imperative to miniaturize optical components, Capasso and his team conceived a revolutionary solution towards the realization of efficient metalenses, transitioning from metallic to dielectric metasurfaces, unveiling a series of breakthroughs, including the discovery of the generalized laws of reflection and refraction [9], and the development of the first efficient flat metalens [10]. This journey was impulsed with a seminal Science paper in 2016, marking a paradigm shift in optics [11], and the birth of a company poised to revolutionize consumer electronic devices with metalenses. The transformative potential of metalenses extends beyond the realm of academia, as collaborations with major semiconductor companies are driving a new era of optical integration in consumer electronics. Through deep-ultraviolet lithography, Capasso and his team demonstrated the feasibility of fabricating metalenses using established semiconductor manufacturing processes, paving the way for their integration into smartphones and other rapidly evolving consumer optical products [12, 13].

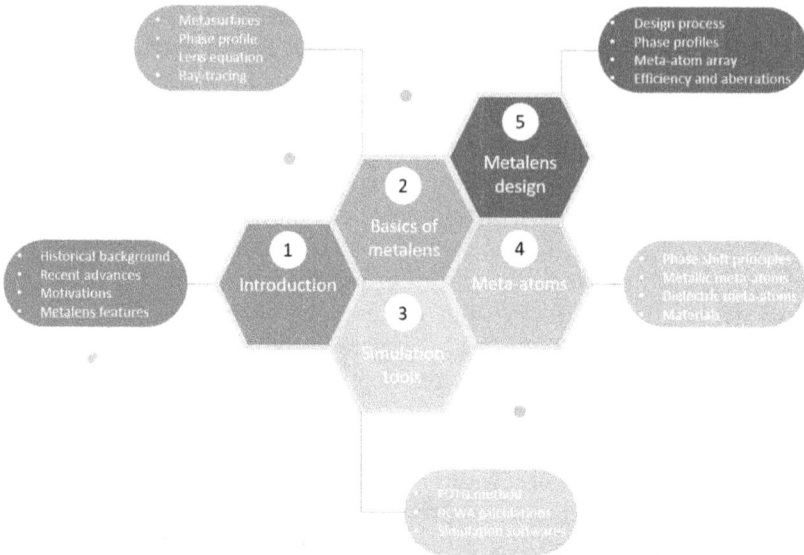

Figure 1.2. In this book, we first introduce metalenses, and then we review the basics of metalens optics, followed by an outlook of the simulation tools, and the optics of meta-atoms. Finally, typical optical design processes are reviewed.

As metalenses continue to evolve, the challenges and opportunities ahead are manifold. From achieving broadband lensing with minimal chromatic aberrations to scaling up metalens, the promise of metalens in reshaping optical technologies across diverse fields, from biomedical imaging to automotive sensor, is undeniable.

In the pages of this introduction to metalens optics (figure 1.2), we embark on a journey through the fundamental optics that define the captivating world of metalenses. Join us to unravel the mysteries of metalens light control and explore the boundless possibilities that lie ahead in the realm of metalens optics.

1.2 Some recent advances

Metalenses, with their ability to manipulate light at the nanoscale, are becoming a revolutionary technology in the field of optics. Recent years have seen a surge of interest and significant advancements in metalens research, driven by innovations in optical designs, nanofabrication techniques, and materials science [14, 15]. These breakthroughs are not only expanding our knowledge and domain of metalenses but also unlocking a plethora of practical applications across various disciplines, ranging from imaging and sensing to polarization manipulation and nonlinear optics. Below, we describe some recent advances.

1.2.1 Large FOV metalens

Imaging is the main ability of any lens for practical applications. There are important optical parameters in imaging systems like field of view (FOV), numerical aperture (NA), and image resolution. FOV is generally limited by a high NA, a problem that was approached by a quadratic metalens to get a wide FOV with a good focusing efficiency [16]. This large FOV metalens was experimentally applied to fingerprint detection, becoming the most compact imaging system for fingerprints to date.

1.2.2 Tunable metalenses

A tunable metalens was designed, fabricated and implemented to realize high-contrast optical sectioning in fluorescence imaging [17]. This is a Moiré metalens that consists of two complementary metasurfaces, with a variable focal length by metasurface rotation. Such tunable metalenses can be applied in microscopy for axial scanning to acquire multiplane images. Because its small size, these metalens systems may find clinical applications in fluorescence microscopy and endoscopy.

1.2.3 Outdoor imaging

Another practical implementation is the first metalens for outdoor imaging in external passive lighting conditions, for imaging under sunlight illumination [18]. Such white light metalens operation is not trivial, and a detailed discussion regarding the limits of achieving broadband imaging is discussed in [19].

1.2.4 Metalens microscopes

The ultrathin nature, alignment by fabrication techniques, phase and polarization controlability, and multifunctionality features make metalens technology perfect for implementing miniature optical systems [20]. For instance, metalenses have been used in compact microscopy systems [21–23]. In microscopy, wide-angle imaging is required, but the FOV is often limited by the off-axis aberrations, in particular, the coma aberration. Then, aplanatic metalenses that correct both spherical and coma aberrations simultaneously are being investigated. In [21] a wide-field microscopy imaging with a FOV of 18° was achieved with a metalens array used to cover a wide section of CMOS image sensors. Exciting applications include miniature two-photon microscopy for investigating brain activity in freely moving animals [22].

1.2.5 Other applications

Another example application in miniature systems is a compact high-resolution spectrometer with a multifocal metalens, where the metalens can simultaneously realize wavelength splitting and light focusing [24]. The compactness and low weight are the main features of metalenses for applications related to augmented and virtual reality [25–27]. An example is the demonstration of a millimeter-scale diameter, high-NA, metalens-based virtual reality system with a home-built fiber scanning near-eye display [25]. Recently, a triple-wavelength metalens was implemented in a compact projection optics for a near-eye display [26], where a large-scale metaoptics was proved in a virtual-reality platform by using a meta-eyepiece and a laser back-illuminated micro-liquid crystal display. In [27] an optical system integrates a polarization-independent metalens with micro light-emitting diodes (LEDs), where

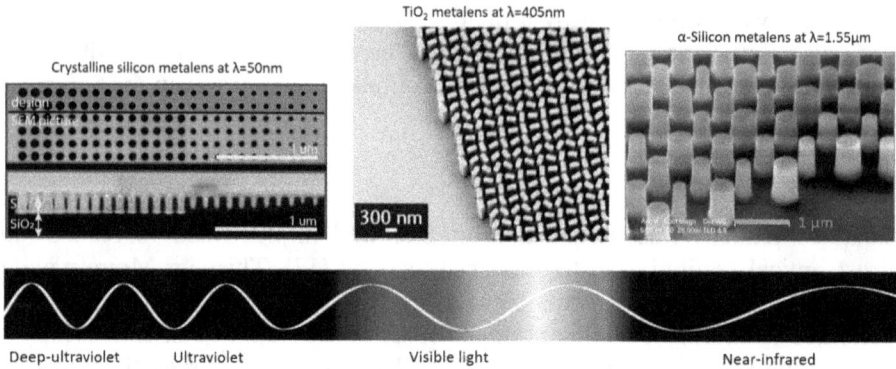

Figure 1.3. Metalenses work at several wavelengths across the electromagnetic spectrum. Insets show a segment of three examples of dielectric metalenses. Materials and operating wavelength λ, of each metalens, are specified at insets. Left inset reproduced with permission from reference [29] CC BY 4.0. Central inset reproduced from reference [11] with permission from AAAS. Right inset reproduced with permission from Springer Nature Customer Service Centre GmbH: [Nature] [30], copyright (2015).

the metalens transforms the light beam of green micro-LEDs into a collimated beam that passes through the eye pupil and images on the retina. In order to make real the application of metalenses in virtual and augmented reality, to become a practically viable option, their optical efficiency needs to be improved, and the chromatic aberration must be addressed.

1.2.6 Across the electromagnetic spectrum

Several metalenses using different materials and operating in various regions of the electromagnetic spectrum [28], ranging from the deep ultraviolet to the infrared have been experimentally demonstrated (figure 1.3). Their performances change depending on the material, and the fabrication constraints that determine the nanostructured surface finishing. The refractive index of the material used determines the degree of light confinement at the operating wavelengths. In addition, the type of nanostructure imparts a phase delay through different mechanisms like resonant effects, propagation, and geometric phase. All of these factors collectively contribute to the differences in optical performance among metalenses.

1.3 Motivations

Metalenses are a significant advancement in the realm of nanotechnology within the ever-changing world of optics, where nanotechnology is revolutionizing economies and societies in the modern era. Offering unprecedented capabilities to manipulate light and imaging, metalenses usher in possibilities beyond the reach of traditional lenses. Unlike their bulky, curved glass counterparts, metalenses boast an advanced, flat design, rendering them lightweight and eminently suitable for very compact spaces and low-weight applications. Moreover, while the fabrication of glass lenses typically requires intricate polishing machinery, metalenses can potentially be assembled using semiconductor chip technology. With their distinctive characteristics, metalenses hold the promise of transforming the landscape for lightweight cameras, mobile phones, and various other devices reliant on lenses for capturing images.

Mobile phone cameras commonly incorporate approximately six transparent refractive lenses. These traditional lenses are made of different materials and shapes, with their special optical design to correct both chromatic and monochromatic aberrations. However, achieving accurate optical alignment for these lenses is a very difficult task that becomes harder as the number of lenses increases. While numerous companies can mass-produce injection-molded plastic lenses, only a select few possess the capability to achieve meticulous alignment with high efficiency for mobile-phone cameras. Although their optical performance is excellent, their weight and volume present limitations for integrating miniaturized optical systems. These technological and manufacturing challenges represent obstacles to further reducing the size of high-performance mobile phone cameras. In contrast, flat metalenses offer the promise of further miniaturization of optical systems. Comprising subwavelength nanostructures with a uniform and thin height profile, metalenses are practically flat. This flatness significantly alleviates the alignment challenge, as alignment techniques widely used in the semiconductor arena can be applied [28].

Lenses are vital optical devices used extensively in imaging, optical communications, vision correction, projectors and displays, and various other applications. With the continuous growth in demand for optical lenses, the potential market for metalenses is immense. Traditional lenses, however, are constrained by fixed volumes and weights, limiting miniaturization. In modern times, refractive lens systems tend to be bulky and expensive. In the pursuit of miniature optical systems, metalenses emerge as the best potential solution. Metalenses offer advantages such as compact structure, easy alignment, lightweight design, multifunctionality, and potentially lower costs. It's no wonder metalenses are currently under extensive investigation.

Numerous approaches to metalenses have already been demonstrated. Their potential applications span across various fields including cellular phone cameras, drone cameras, micro projectors, machine vision, eyeglasses, virtual and augmented reality, and lab-on-a-chip devices, among others. There is a big chance that flat optics will play an increasingly important role in all areas of optical technology.

Metalenses possess several distinctive characteristics and features that make them a compelling subject of study and innovation. One of these attributes is their capability to focus light to subwavelength spots with high efficiency. This property holds immense potential for pushing the boundaries of small and compact optical systems. Multifunctionality is another attribute, as metalenses exhibit multifunctional behavior, adapting their optical properties based on various degrees of freedom, such as wavelength, polarization, and incident angle. Additionally, their unique ultrathin planar form factor is remarkable. The planar, ultrathin nature of metalenses leads to compact and lightweight optical systems. This feature is particularly advantageous for applications where space and weight constraints are critical considerations. Scalability of manufacturing is another attribute. Metalenses can be manufactured using scalable microfabrication semiconductor techniques used in microchip fabrication, offering a pathway to cost-effective production. This scalability is essential for integrating metalenses into a variety of optical devices and systems. In the following section we detail deeper these distinctive characteristics and features of metalenses.

1.4 Features and potential advantages of metalenses

Metalenses have special optical characteristics, offering several potential advantages over traditional diffractive and refractive lenses (figure 1.4). Unlike their bulky counterparts, metalenses harness the power of metasurfaces to achieve compact and multifunctional systems, and in general allow the implementation of miniature optical systems. These characteristics redefine the landscape of optical design and pave the way for a new era of miniaturized optics. Metalenses exhibit several distinctive features, including subwavelength focusing with high efficiency, multifunctionality, tunability, ultrathin planar form factor, and great potential for low-cost manufacturing. Below, we describe some of their properties and potential advantages [15, 31].

1.4.1 Subwavelength wavefront control

At the heart of metalenses lies the ability to manipulate light at the subwavelength scale, thanks to meticulously engineered nanostructures. Their nanoelements induce precise phase delays across the optical wavefront, enabling unprecedented control over light propagation and focusing. Unlike classical lenses, which rely on large structural features, metalenses achieve their remarkable performance through subwavelength spaced, quasiperiodic structures.

1.4.2 Reduced thickness and binary structure

Metalenses offer a significant advantage in terms of thickness and structure. With characteristic thicknesses on the order of $\lambda/10$ to λ nm, limited only by the substrate

Figure 1.4. Metalenses among other lens types. (a) Conventional monolithic refractive lens. (b) Diffractive lens with its radial zones. (c) Metalens with its 2D pattern of nanoelements. Reproduced with permission from reference [31].

thickness (typically ~1 mm thick), metalenses not only enhance miniaturization but also simplify manufacturing processes, making them promising candidates for next-generation optical devices. Their reduced thickness makes them ideal for assembling meta-eyepieces in virtual and augmented reality applications, where compactness and lightweight are precious properties.

Its flatness confines the metalens structure to a binary pattern (binary: two-level surface height), which is considered easier to manufacture compared to multilevel diffractive lens (surface profile consisting of several discrete surface heights). Moreover, this binary structure offers unlimited optical design freedom at the nanoscale, challenging nanotechnology, and modernizing lenses the last analog bastion of optics with the development of the first digital lens, the metalens.

1.4.3 CMOS fabrication compatibility

The fabrication of metalenses is highly compatible with the complementary metal-oxide-semiconductor (CMOS) fabrication process in the mature microelectronics industry [20]. The compatibility of metalenses with CMOS fabrication processes further enhances their appeal, enabling seamless integration into modern light sensors and on-chip systems. Metalenses could be directly integrated with camera image sensors, and advancements in semiconductor foundries could allow precise alignment of optical components, potentially replacing conventional systems with cascading lenses. However, particular challenges for further technological development of metalens-based integrated systems must be addressed. For example, substrate materials must be CMOS-compatible (e.g., Si) and be lossless in the visible and infrared wavelengths, to reduce the difficulties of metalens integration, preparation, and encapsulation. There are several CMOS processing technologies under research and development [32]. This CMOS compatibility will enable high-volume manufacturing of metalens commercial applications ranging from smartphone cameras and IoT (Internet of Things) to automotive sensors.

1.4.4 Tunability

Recent advancements have showcased the tunability of metalenses, allowing for dynamic focus adjustment with a performance comparable with that of typical mechanical tunable refractive lenses [33]. Tunable metalenses based on different technologies are under research, for example graphene-based metalenses with electrically tuneable focusing. Also tunable or zooming metalenses on a stretchable substrate have been developed. Other mechanical tunability mechanisms such as Alvarez metalenses and metalenses with MEMS systems have been experimentally demonstrated. The simplest approach, the Alvarez metalens, is a zoom lens composed of two special metasurfaces that make zoom by controlling the relative displacement between the two metasurfaces. There are several tunable metalens technologies under development. Their tunable properties set metalenses apart, offering versatility and adaptability beyond the capabilities of classical lenses.

1.4.5 Polarization selectivity

Metalenses can introduce polarization selectivity, enabling tailored functionalities for specific applications [15]. Unlike classical lenses, which exhibit uniform behavior across polarizations, some metalenses focus differently polarized light at distinct positions, opening avenues for polarization-sensitive optics. The polarization selectivity of certain metalens desings allows full-Stokes imaging polarimetry. Full-Stokes metalens polarization cameras have been demonstrated. The metasurface area is splitted into different polarization bases (set of reference states that allows to describe any arbitrary polarization state), and all Stokes' parameters are measured at the focal plane, which is divided into superpixels. In other words, the segmented metalens splits light using polarization bases of linear and left-/right-handed circular polarizations, and focus them on different sections of the image sensor, thus directly measuring all Stokes' parameters with high resolution. This is convenient because traditional imaging polarimeters filter light using four polarizers and provide only three of the four Stokes' parameters. However, the metalens imaging polarimeter can measure all Stokes' parameters with higher signal-to-noise ratio.

1.4.6 High-numerical aperture capability

Metalenses have demonstrated high numerical aperture (NA) capability. Although this was previously attainable with conventional lenses, a refractive lens becomes fatter as NA increases, making it impractical. Therefore, metalenses promise high NA with compactness and miniaturization. Metalenses with high NA can achieve the same or even better optical performance compared to traditional lenses while being much thinner and lighter. This compactness is advantageous for various applications, including integrated photonics and wearable devices. High NA is particularly important in applications such as microscopy. Also, a high NA allows metalenses to collect light from a wider range of angles, increasing the amount of light that can be focused onto the image plane. This not only leads to brighter and sharper images but is also attractive for non-imaging optics [34].

1.4.7 Multifunctional capability

One of the most prominent features of metalenses is their multifunctional capability [15, 20]. For some advanced applications such as polarization imaging and 3D imaging, simultaneous or tunable multi-foci are required. The multifunctionality capability of metalenses can avoid cascading several optical elements with different functionalities, without added mechanical components. Metalenses can function differently based on various degrees of freedom of light, including wavelength, polarization, and incident angle. For instance, metalenses composed of nanopost meta-atoms with asymmetric cross-sections can impart arbitrary wavefronts on orthogonal polarizations of light. In such a basic example of multifunctionality metalenses can impose two independent phase profiles for two different orthogonal polarizations by using meta-atoms with mirror symmetry. For example, a bifunctional metalens can focus right-handed and left-handed polarized light into a simple spot and a doughnut-shaped spot, respectively.

Similarly, wavelength degree of freedom enables the realization of multi-wavelength metalenses, facilitating applications requiring focusing at discrete wavelengths. Their multifunctionality is enabling their incorporation in advanced optical applications like computational imaging and virtual and augmented reality.

In summary, metalenses represent a paradigm shift in optical design, offering unparalleled performance and versatility. While all type of lenses have their merits, the unique characteristics of metalenses propel them to the forefront of optical innovation. As the field continues to evolve, metalenses are poised to revolutionize a myriad of applications, from imaging and spectroscopy to augmented reality. Their journey from academic research to commercialization marks a pivotal moment in the history of optics, promising transformative advancements in the years to come.

Bibliography

[1] Hecht E 2002 *Optics* (San Francisco, CA: Addison-Wesly)

[2] Lalanne P and Chavel P 2017 Metalenses at visible wavelengths: past, present, perspectives *Laser Photonics Rev.* **11** 1600295

[3] Khorasaninejad M and Capasso F 2017 Metalenses: versatile multifunctional photonic components *Science* **358** eaam8100

[4] Huang F M, Zheludev N, Chen Y and Javier Garcia de Abajo F 2007 Focusing of light by a nanohole array *Appl. Phys. Lett.* **90** 091119

[5] Sun Z and Kim H K 2004 Refractive transmission of light and beam shaping with metallic nano-optic lenses *Appl. Phys. Lett.* **85** 642–4

[6] Shi H *et al* 2005 Beam manipulating by metallic nano-slits with variant widths *Opt. Express* **13** 6815–20

[7] Verslegers L *et al* 2009 Planar lenses based on nanoscale slit arrays in a metallic film *Nano Lett.* **9** 235–8

[8] Quidant R, Aumiller W and Capasso F 2024 An interview with Federico Capasso *ACS Photonics* **11** 811–5

[9] Yu N *et al* 2011 Light propagation with phase discontinuities: generalized laws of reflection and refraction *Science* **334** 333

[10] Aieta F, Genevet P, Kats M A, Yu N, Blanchard R, Gaburro Z and Capasso F 2012 Aberration-free ultrathin flat lenses and axicons at telecom wavelengths based on plasmonic metasurfaces *Nano Lett.* **12** 4932–6

[11] Khorasaninejad M *et al* 2016 Metalenses at visible wavelengths: diffraction-limited focusing and subwavelength resolution imaging *Science* **352** 1190–4

[12] She A, Zhang S, Shian S, Clarke D R and Capasso F 2018 Large area metalenses: design, characterization, and mass manufacturing *Opt. Express* **26** 1573–85

[13] Park J-S, Zhang S, She A, Chen W T, Lin P, Yousef K M A, Cheng J-X and Capasso F 2019 All-glass, large metalens at visible wavelength using deep-ultraviolet projection lithography *Nano Lett.* **19** 8673–82

[14] Kuznetsov A I *et al* 2024 Roadmap for optical metasurfaces *ACS Photonics* **11** 816–65

[15] Arbabi B and Faraon A 2023 Advances in optical metalenses *Nat. Photon* **17** 16–25

[16] Lassalle E, Mass T W, Eschimese D, Baranikov A V, Khaidarov E, Li S, Paniagua-Dominguez R and Kuznetsov A I 2021 Imaging properties of large field-of-view quadratic metalenses and their applications to fingerprint detection *ACS Photonics* **8** 1457–146

[17] Luo Y *et al* 2021 Varifocal metalens for optical sectioning fluorescence microscopy *Nano Lett.* **21** 5133–42

[18] Engelberg J, Zhou C, Mazurski N, Bar-David J, Kristensen A and Levy U 2020 Near-IR wide-field-of-view huygens metalens for outdoor imaging applications *Nanophotonics* **9** 361–70

[19] Engelberg J and Levy U 2021 Achromatic flat lens performance limits *Optica* **8** 834–45

[20] Pan M, Fu Y, Zheng M, Chen H, Zang Y, Duan H *et al* 2022 Dielectric metalens for miniaturized imaging systems: progress and challenges *Light: Sci. Appl.* **11** 195

[21] Xu B B *et al* 2020 Metalens-integrated compact imaging devices for wide-field microscopy *Adv. Photonics* **2** 066004

[22] Wang C *et al* 2023 Miniature two-photon microscopic imaging using dielectric metalens *Nano Lett.* **23** 8256–63

[23] Li Z X *et al* 2021 Compact metalens-based integrated imaging devices for near-infrared microscopy *Opt. Express* **29** 27041–7

[24] Pahlevaninezhad H *et al* 2018 Nano-optic endoscope for high-resolution optical coherence tomography *in vivo Nat. Photon.* **12** 540–7

[25] Li Z *et al* 2021 Meta-optics achieves RGB-achromatic focusing for virtual reality *Sci. Adv.* **7** eabe4458

[26] Li Z *et al* 2022 Inverse design enables large-scale high-performance meta-optics reshaping virtual reality *Nat. Commun.* **13** 2409

[27] Li S-H *et al* 2024 Augmented reality system based on the integration of polarization-independent metalens and micro-LEDs *Opt. Express* **32** 11463–73

[28] Chen W T, Zhu A Y and Capasso F 2020 Flat optics with dispersion-engineered metasurfaces *Nat. Rev. Mater.* **5** 604–20

[29] Ossiander M, Meretska M L, Hampel H K, Lim S W D, Knefz N, Jauk T and Schultze M 2023 Extreme ultraviolet metalens by vacuum guiding *Science* **380** 59–63

[30] Arbabi A, Horie Y, Ball A J, Bagheri M and Faraon A 2015 Subwavelength- thick lenses with high numerical apertures and large efficiency based on highcontrast transmitarrays *Nat. Commun.* **6** 7069

[31] Engelberg J and Levy U 2020 The advantages of metalenses over diffractive lenses *Nat. Commun.* **11** 1991

[32] Li N, Xu Z, Dong Y, Hu T, Zhong Q, Fu Y H, Zhu S and Singh N 2020 Large-area metasurface on cmos-compatible fabrication platform: driving flat optics from lab to fab *Nanophotonics* **9** 3071–87

[33] Chen M K, Wu Y, Feng L, Fan Q, Lu M, Xu T and Tsai D P 2021 Principles, functions, and applications of optical meta-lens *Adv. Opt. Mater.* **9** 2001414

[34] Moreno I, Avendaño-Alejo M and Castañeda-Almanza C P 2020 Nonimaging metaoptics *Opt. Lett.* **45** 2744–7

Chapter 2

Basics of metalens optics

A metalens is a metasurface with unique focusing and imaging characteristics. This special metasurface is an array of nano-spaced optical elements (also known as meta-atoms, nano-scatterers, nanoelements, nano-antennas, or nanobricks) at a flat surface, whose primary function is to transform a plane wavefront into a spherical wavefront. Every meta-atom locally shifts the phase of an incident wavefront of light [1, 2]. And because a wavefront is the surface or locus of points that have the same phase [3], a metalens offers the fascinating ability to shape the wavefront of the transmitted light into an spherical wavefront due to its radial array of meta-atoms. This wavefront transformation is produced by radially adjusting the phase shift introduced by each of the nanoelements. The meta-atom phase shift can be adjusted by changing its geometrical parameters (such as size, shape, or orientation across the flat surface). All these metalens characteristics may be described from the viewpoint of basic optics. For this purpose, this chapter introduces metasurfaces, phase profiles, meta-atoms, and paraxial metalens optics.

2.1 Optics of metasurfaces

In the fascinating realm of flat optics, a metasurface is like a Lego construction set, with meta-atoms as building bricks that are used for building a two-dimensional world of optical devices. Metasurfaces often comprise thousands or millions of meta-atoms of certain materials and shapes arranged to locally: shift the wavefront phase, suppress or enhance reflectance-transmittance, or to alter the polarization characteristics of light. A metasurface can be a periodic array of nano-structures replicated in a one or two-dimensional arrangement of different nanoelements. The meta-atoms may be dielectric and/or metallic structures. These building blocks include sets of nanorods, cylindrical holes, nanofins, V-shaped antennas, and so on.

Light waves, whose nature is periodic, interact with a metasurface in a unique way because the scale of the meta-atoms' periodicity is a fraction of the light wavelength. Additional degrees of freedom emerge because light waves can be shifted in phase without accumulating length propagation through a medium. Meta-atoms introduce an abrupt phase change to waves under reflection or transmission, called phase shift [4], behaving in a manner akin to optical path length, but applicable to flat optics. In addition, the subwavelength periodicity removes the formation of diffraction orders, which is characteristic of conventional diffractive optics, where the periodicity of optical elements is of the same order as that of the wavelength. In metalenses, the absence of diffraction orders improves the efficiency by removing virtual focal spots [1].

2.1.1 The law of refraction–reflection for metasurfaces

When a ray of light is traveling through a metasurface, part of the energy is reflected and part transmits the metasurface. The reflected and transmitted rays are bent at the boundary. The direction of refracted and reflected light depends on the angle of incidence, and on the spatial variation of the metasurface phase profile. This surface phase profile has a phase gradient $\nabla\Phi$ that modifies the traditional Snell law of refraction. The generalized law of refraction/reflection of a metasurface that has radial phase variation $\Phi(r)$, is [4]:

$$\frac{1}{k_o}\frac{d\Phi}{dr} = n_2 \sin\theta_2 - n_1 \sin\theta_1, \tag{2.1}$$

where θ_1, θ_2 are the angles of incidence and of refraction/reflection, respectively. Also, n_1 and n_2 are the refractive indices of the medium where light is incident and transmitted/reflected, respectively; and k_o is the wave number $(2\pi/\lambda_o)$, where λ_o is the vacuum wavelength. If the phase profile Φ is constant $d\Phi/dr = 0$, then equation (2.1) becomes the well-known Snell law (or the reflection with $n_1 = n_2$). The reflection law is obtained by making n_2 equal to n_1.

2.1.2 Derivation of the refraction law from Huygens's principle

The law of refraction/reflection of metasurfaces was first derived from Fermat's principle and experimentally confirmed in a fundamental research paper [4], and recently was derived by using Huygens's principle [5]. Let us present the Huygens's principle approach due to it is particularly illustrative procedure.

According to Huygens's principle, a plane wavefront propagates as if it were composed of many point sources, each emitting a spherical wave, such that the wavefront at some later time is the envelope of these wavelets. A metasurface introduces different phase shifts at every point on it, which is equivalent to introducing different time retardations across the metasurface. This behavior is illustrated in figure 2.1(a), which shows a wavefront striking a metasurface between two dielectric materials of indices n_1 and n_2. The transmitted wavefront is represented by the line DC, and the line AB represents the incident wavefront, just as it strikes the metasurface at point A. At that instant, the wave at A is retarded a short time due to the metasurface in that point (green point), and then it sends out a Huygens's wavelet toward D. At the same time, the wave at B emits a Huygens's wavelet toward C, and after a short time retardation in the metasurface (blue point)

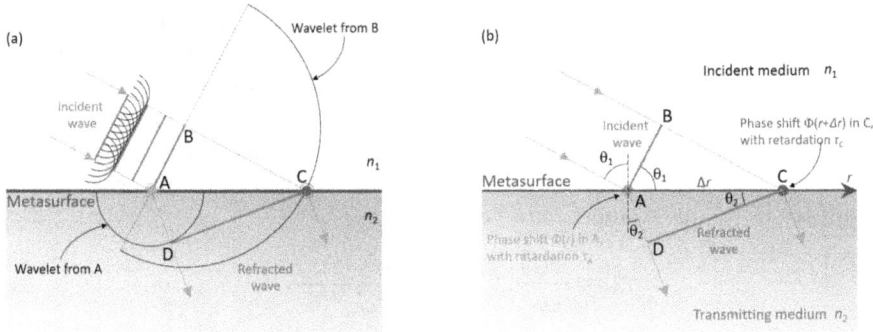

Figure 2.1. Refraction of a wave at a metasurface, diagrams with Huygens's construction for deriving the law of refraction. (a) Showing wavelets, and (b) showing angles and phase shifts. Consider a metasurface surrounded by two media with indices n_1 and n_2. This metasurface introduces a phase shift in function of its phase profile $\Phi(r)$, which shifts time or delays the wave. Light that hits point A is retarded there (green point), then travels to point D, and in the same time interval, light moves from B to C, and is retarded there (blue point). In other words, at the instant that a light ray strikes the metasurface, its phase is shifted by Φ or equivalently it is retarded by τ. This time delay τ must be included in time of travel of the light from A to D, and from B to C. (b) Reproduced from [5], copyright (2022) European Physical Society.

it crosses the metasurface. This behavior is summarized in figure 2.1(b), which depicts the wavefront at two instants of time, AB at the instant of incidence and CD after the refraction. In other words, light that hits A is retarded there, then travels to D, and in the same time interval, light moves from B to C, and is retarded there.

According to the Huygens's principle, these time intervals must be the same for all the points of a wavefront, and by equating the time of light travel from A to D and B to C inside the transmitting medium, we obtain the relation:

$$t_{AD} + \tau_A = t_{BC} + \tau_C, \tag{2.2}$$

where t_{BC} is the time it takes for point B on the wavefront (traveling at speed v_1) to reach point C, and t_{AD} is the time it takes the transmitted portion of that same wavefront (traveling at speed v_2) to reach point D from point A. The time delays due to the metasurface at points A and C are τ_A and τ_C, respectively.

Let us deduce the four elements in equation (2.2). Traveling times of the light are $t_{AD} = AD/v_2$ and $t_{BC} = BC/v_1$, where AD is the distance from A to D, and BC is the distance between B and C. Noting that the two right triangles ADC and ABD share a common hypotenuse (AC $= \Delta r$), it is easy to find that $t_{AD} = \Delta r\,\sin\theta_2/v_2$ and $t_{BC} = \Delta r\,\sin\theta_1/v_1$. By using the definition of refractive index of a medium ($n = c/v$, where c is the speed of light in vacuum), we can write:

$$t_{AD} = \frac{1}{c}\Delta r\; n_2 \sin\theta_2, \quad t_{BC} = \frac{1}{c}\Delta r\; n_1 \sin\theta_1, \tag{2.3}$$

The other terms of equation (2.2) are the time retardations τ_A and τ_C, introduced by the metasurface. The light transmitted through the point A in the metasurface is phase shifted by Φ_A, which is indeed equivalent to introduce a time delay τ_A in the transmitted beam. Remember that the phase of a wavelet is the argument ($kr - \omega t$)

of the wavefunction, where the temporal part of the phase is ωt, where ω is the wave frequency ($\omega = ck_o$). Therefore, a phase shift in point A affects the temporal part, and it is equivalent to a time retardation, given by $\Phi_A = \omega\tau_A = ck_o\tau_A$. The same is true for the point C, where $\Phi_C = \omega\tau_C = ck_o\tau_C$. Considering that A is located at the coordinate r, and that C is in the coordinate $r + \Delta r$, the phase shifts can be written as $\Phi_A = \Phi(r)$ and $\Phi_C = \Phi(r + \Delta r)$. Then, the retardations introduced by the metasurface in points A and C can be written as

$$\tau_A = \frac{\Phi(r)}{ck_o}, \quad \tau_C = \frac{\Phi(r + \Delta r)}{ck_o}, \tag{2.4}$$

Finally, substituting equations (2.4) and (2.3) into equation (2.2), canceling c, and rearranging terms, gives

$$\frac{1}{k_o}\left[\frac{\Phi(r + \Delta r) - \Phi(r)}{\Delta r}\right] = n_2 \sin\theta_2 - n_1 \sin\theta_1. \tag{2.5}$$

This equation is evaluated at two points, but equation (2.1) is happening at a single point in space, i.e., $\Delta r = 0$. In the limit as Δr approaches 0, the two parts of the wavefront will become infinitely close to each other. Therefore, the left side of equation (2.5) becomes a derivative $d\Phi/dr$ of the phase Φ in the limit $\Delta r \to 0$, and then equation (2.5) becomes equation (2.1). This is a derivation of the refraction law of metasurfaces by using Huygens's principle. Note that this result is valid for light reflection if we consider θ_2 as the angle of reflection, and the refractive index $n_2 = n_1$.

2.1.3 The 3D law of refraction–reflection for metasurfaces

In general the metasurface may have an arbitrary phase variation $\Phi(x,y)$. In this case, the generalized refraction–reflection law at an interface with a 2D phase gradient is given by two coupled equations. By considering that the direction of propagation of the incident light beam is given by (θ_1, φ_1), and the metasurface oriented as seen in figure 2.2, the generalized law of refraction–reflection is [6, 7]:

$$\begin{cases} \dfrac{1}{k_o}\dfrac{\partial\Phi}{\partial x} = n_2 \sin\theta_2 \cos\varphi_2 - n_1 \sin\theta_1 \cos\varphi_1 \\ \dfrac{1}{k_o}\dfrac{\partial\Phi}{\partial y} = n_2 \sin\theta_2 \sin\varphi_2 - n_1 \sin\theta_1 \sin\varphi_1 \end{cases}, \tag{2.6}$$

where n_1 and n_2 are the refractive indices of the medium where light is incident and transmitted/reflected, respectively, and k_o is the wave number ($2\pi/\lambda_o$), where λ_o is the vacuum wavelength. Angle φ_2 is the azimuthal angle of refraction/reflection in the plane xy (metasurface plane), and φ_1 is the azimuthal angle of incidence. Angle θ_1 is the altitude angle of incidence with respect to the z-axis (normal to the metasurface plane), and θ_2 is the altitude angle of the light refracted (or reflected). Equation (2.6) shows that even if the incident ray is normal to the metasurface (xy plane), $\theta_1 = 0$, the transmitted/reflected ray may lie out of the z-axis, if there is a 2D phase gradient. It is therefore not possible to define a plane of incidence, the one that contains the

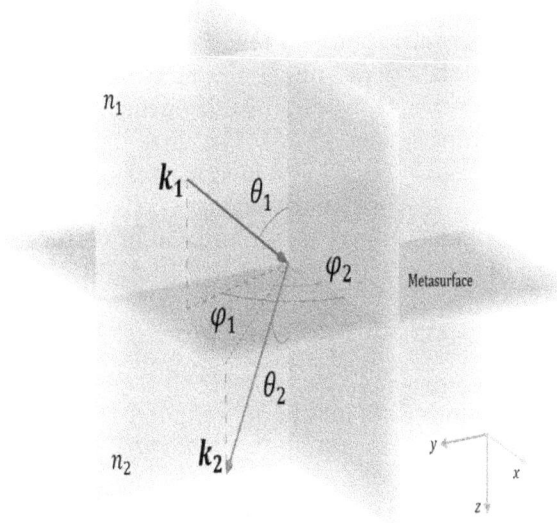

Figure 2.2. Diagram of the generalized refraction at a metasurface with arbitrary 2D phase profile $\Phi(x,y)$. The incident light ray has a normalized wave-vector k_1 with arbitrary direction. The refracted ray acquires a direction given by a unitary vector k_2. The metasurface is in the xy plane at the interface between two media with refractive indices n_1 and n_2. Reprinted with permission from [6]. Copyright 2022 Optica Publishing Group.

incident, the reflected and the transmitted beams. Note that when the phase gradient is oriented along the x-axis ($\partial\Phi/\partial y = 0$), and the incident light is on the xz plane ($\varphi_1 = 0$) the transmitted/reflected beams are in plane ($\varphi_2 = 0$), and equation (2.6) becomes equation (2.1).

Equation (2.6) is a generalized reflection and refraction law in 3D that accounts for any orientation of phase discontinuity gradients or 2D phase gradients at a metasurface. This law in 3D was first deduced and experimentally demonstrated in plasmonic metasurfaces [8]. While equation (2.6) may appear different from the one provided in reference [8], they are in fact equal. The disparity arises due to certain assumptions and the use of a different spherical coordinate system, in reference [8].

The 3D law of refraction–reflection for metasurfaces can be expressed for a polar phase gradient $\nabla\Phi = \left(\frac{\partial\Phi}{\partial r}, \frac{\partial\Phi}{\partial\psi}\right)$. Let us assume that the phase profile $\Phi(r, \psi)$ is a continuous function of the radial position r on the metasurface, and of the polar angle ψ. After some mathematical treatments equation (2.6) can be written in polar coordinates as [6]:

$$\begin{cases} \cos(\psi)\dfrac{\partial\Phi}{\partial r} - \dfrac{1}{r}\mathrm{sen}(\psi)\dfrac{\partial\Phi}{\partial\psi} = k_0 n_2 \sin\theta_2 \cos\varphi_2 - k_0 n_1 \sin\theta_1 \cos\varphi_1 \\[2mm] \mathrm{sen}(\psi)\dfrac{\partial\Phi}{\partial r} + \dfrac{1}{r}\cos(\psi)\dfrac{\partial\Phi}{\partial\psi} = k_0 n_2 \sin\theta_2 \sin\varphi_2 - k_0 n_1 \sin\theta_1 \sin\varphi_1 \end{cases} \quad (2.7)$$

where $r = \sqrt{x^2 + y^2}$.

2.1.4 The 3D vector form of refraction–reflection law

In metasurfaces, the beam's refraction and reflection direction is determined by the phase spatial profile. Equation (2.6) determines the direction of transmitted or reflected beam through a 2D phase profile. However, ray tracing methods through metasurfaces may require a vector form of equation (2.6). The three-dimensional vector form of 'Snell's' law for metasurfaces can be deduced by using a geometric approach [6].

First note that equation (2.6) depends on the x-axis and y-axis phase profile gradients. This 2D phase gradient may be represented in vector form $\nabla\Phi$ in the xy plane of the metasurface (figure 2.3), given by

$$\nabla\Phi = \frac{\partial\Phi}{\partial x}\hat{i} + \frac{\partial\Phi}{\partial y}\hat{j}. \tag{2.8}$$

Now let us combine equations (2.8) and (2.6), and after rearranging terms, we can write:

$$\frac{1}{k_0}\nabla\Phi = n_2(\sin\theta_2\cos\varphi_2\hat{i} + \sin\theta_2\sin\varphi_2\hat{j})$$
$$- n_1(\sin\theta_1\cos\varphi_1\hat{i} + \sin\theta_1\sin\varphi_1\hat{j}). \tag{2.9}$$

Considering the geometry shown in figure 2.2, the direction of propagation of the incident light beam and the transmitted (or reflected) one are given by the normalized wavevectors k_1 and k_2, respectively. These wavevectors may be expressed as function of the angles $\theta_{1(2)}$ (angle between $k_{1(2)}$ and z-axis), and $\varphi_{1(2)}$ (the azimuthal angle formed by the x-axis and the projection of $k_{1(2)}$ on the xy plane). With this choice of coordinates, the normalized wavevectors are:

$$k_{1(2)} = \begin{pmatrix} \sin\theta_{1(2)}\cos\varphi_{1(2)} \\ \sin\theta_{1(2)}\sin\varphi_{1(2)} \\ \cos\theta_{1(2)} \end{pmatrix}. \tag{2.10}$$

Here the subindex '1' represents the incident light, and the subindex '2' indicates the transmitted or reflected beam. By using equation (2.10) we can identify the k components along the x–y–z-axes in equation (2.9), and by using the definition of cross product in equation (2.9), the refraction–reflection law of metasurfaces may be written in vector form as:

$$n_2(n \times k_2) \times n - n_1(n \times k_1) \times n = \frac{1}{k_0}\nabla\Phi. \tag{2.11}$$

where k_1 and k_2 are the unit vectors along the directions of the incident ray and refracted (or reflected) ray, and n is the unit vector along the normal to the metasurface. The equation (2.11) is valid for any normal n because the cross product is invariant under rotations of the coordinate system. Equation (2.11) reduces to the vector form of classical Snell law in purely refractive interfaces by doing $\nabla\Phi = 0$ (figure 2.3).

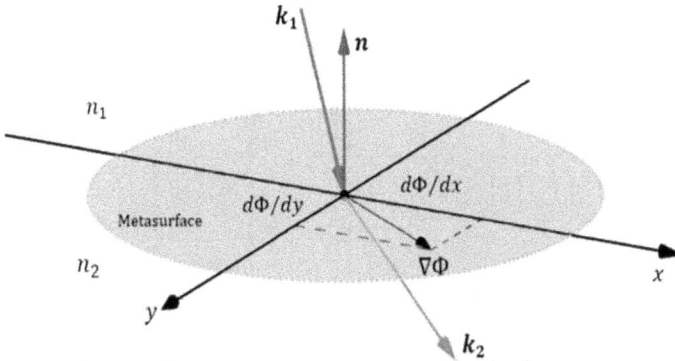

Figure 2.3. Scheme of the refraction showing the 2D phase gradient vector on a metasurface. Also it is shown the unit vectors of incident beam k_1 and the refracted beam k_2, and n the normal vector of the metasurface. The phase gradient has two components, $\partial\Phi/\partial x$ and $\partial\Phi/\partial y$ at each point (x, y) on the metasurface. Reprinted with permission from [6]. Copyright 2022 Optica Publishing Group.

2.2 Metalens phase profile

One could say that the phase profile is to a metalens what the spherical shape is to a glass lens. A metalens is a flat lens comprising a metasurface or nanostructure with circular symmetry, precisely engineered to generate a radial phase profile $\Phi(r)$. This phase profile is required to make spherical wavefronts, in order to focus incoming plane wavefronts to a point. The phase profile of metalenses is fundamental to their focusing and imaging functions. Unlike traditional lenses, metalenses utilize precisely engineered metasurfaces to manipulate the phase of incoming light waves. This phase manipulation enables metalenses to focus light. The phase profile of a metalens can be tailored to achieve specific optical properties, such as focusing, aberration correction, and even multifocal capabilities. Metalenses can exhibit various types of phase profiles tailored to specific applications. These profiles include radial, azimuthal, and even complex multifocal patterns. Radial hyperbolic profiles concentrate light towards a central point, akin to traditional lenses, while azimuthal profiles, for example the sum of radial and spiral phases, may perform edge-enhanced imaging [9]. Multifocal profiles, on the other hand, allow for simultaneous focusing at multiple points. The versatility in phase profile design empowers metalenses to revolutionize fields such as imaging, telecommunications, and microscopy, promising a new era of compact and high-performance optical devices. Now, let us deduce the basic equation of the metalens phase profile, the hyperbolic phase distribution, which can be derived by several methods. This deduction is performed by placing a condition for shaping a plane wavefront into a spherical wavefront. One deduction is by using the law of refraction of metasurfaces equation (2.1), and other is by the geometry and the Huygens's principle.

2.2.1 Derivation of the metalens phase profile from the refraction law

Let us deduce the hyperbolic phase profile by using equation (2.1). For this purpose consider figure 2.4(a), which shows a metalens represented by a nanostructure on a

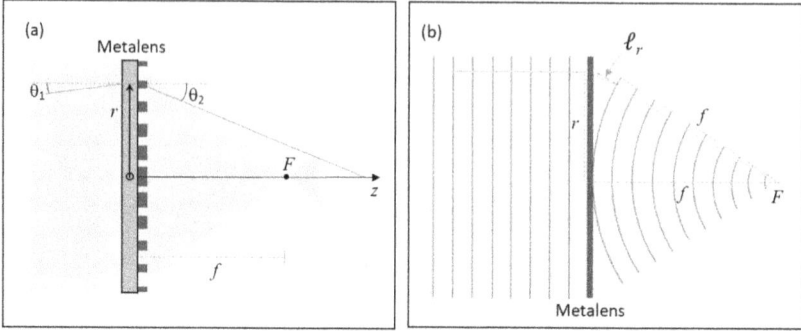

Figure 2.4. (a) Parallel light rays refract in a metalens through the focal point F. (b) Optical path difference ℓ_r introduced by the phase profile of the metalens, to focus light at F. Figure reproduced from [5], copyright (2022) European Physical Society.

transparent plate. An incident plane wavefront is focused through a focal point F. The metasurface must have a phase $\Phi(x,y)$ that makes any incident ray (parallel to the optical axis) at radial distance r, be focused at distance $z = f$ from the metalens (figure 2.4(a)). To make any incident ray to be parallel to the optical axis is equivalent to make $\sin\theta_1 = 0$. To focus this ray along the z-axis at distance f after being transmitted is equivalent to make $\sin\theta_2 = -r(r^2+f^2)^{-0.5}$, where r is the radial distance on the metalens. Note that the sign of $\sin\theta_2$ is negative, which is because θ_2 is positive at the opposite side of the normal (dotted line in figure 2.4(a)) with respect to θ_1, and negative at the same side of the normal, which is the classical convention of the refraction law. By substituting these geometrical conditions for focusing a light beam at a distance f in equation (2.1), a simple differential equation $d\Phi/dr = -rk_0n_2(r^2+f^2)^{-0.5}$ is obtained, whose solution gives the well-known hyperbolic phase profile:

$$\Phi(r) = \pm k\left(\sqrt{r^2 + f^2} - f\right), \tag{2.12}$$

where r is the radial position on the metalens, f is the metalens focal length, and $k = k_0n_2$ is the wave number (and $k_0 = 2\pi/\lambda_0$). The plus or minus sign in equation (2.12) indicates a diverging or converging lens, respectively. Note that equation (2.12) does not depend on n_1, and it only depends on n_2 because the transmitted light is oblique (focused) at medium 2, but the incident light in the medium 1 is not oblique ($\theta_1 = 0$).

2.2.2 Phase profile derivation from geometry and the Huygens's principle

The derivation of the hyperbolic phase profile, equation (2.12), may be achieved from geometry and the Huygens's principle. For this purpose consider that a metalens shapes a plane wavefront into a spherical wavefront (figure 2.4(b)). For performing this transformation, the metalens must introduce a phase shift $\Phi(r)$ that increases radially from the metalens center to its periphery. At every radial distance r, a metalens must change the phase of an incident plane wavefront into a spherical wavefront. From figure 2.4(b), we can see that the transmitted wavefront bends a distance ℓ_r that increases with the radial distance r. Then, the metalens phase profile must transform the optical path by adding an optical path difference ℓ_r for each incident ray at different radial distances r, considering that any optical path difference OPD is equivalent to a phase shift kOPD [3]. Therefore, the required phase shift must be $\Phi(r) = \pm k\ell_r$, which by observing the geometry of figure 2.4(b) is the equation (2.12), where $\ell_r = (r^2 + f^2)^{0.5} - f$.

2.2.3 About the gradient of phase profile

Equation (2.12) is basic for metalenses design; most designs incorporate this phase profile as a first approach, and then it could be called the metalens makers' equation. Figure 2.5(a) shows the hyperbolic phase profile given by equation (2.12), which enables a metalens to focus collimated light into its focal point F. One important aspect regarding the metalens phase profile is that what matters is the spatial variation of the phase or the relative phase, rather than the absolute value of the phase. This is because the physical condition for bending light is the phase gradient $d\Phi/dr$, not the phase Φ itself, as determined by the refraction law (equation (2.1)). For example, one can note from equation (2.12) that the phase along the optical axis is zero, but its on-axis value could be different, i.e. the metalens phase profile can be Φ or $\Phi + A$, where A is an arbitrary constant.

2.2.4 From phase profile to metasurface profile

The metalens phase profile $\Phi(r)$ can be reproduced by radially varying (across its flat surface) the geometrical parameters of nanoelements, such as size, shape, or orientation. Except for the nanofins, where the phase shift is twice their angle of rotation, there is not a unique relationship between these geometrical parameters and the phase shift the meta-atoms introduce. The phase shift introduced by meta-atoms depends on many parameters like its refractive index, its height, its absorption, the unitary cell size, the wavelength, etc. Although some semi-analytical models have been reported [10], there is no a single mathematical relation or analytic deduction between Φ and the size or shape of meta-atoms. Such a relationship is

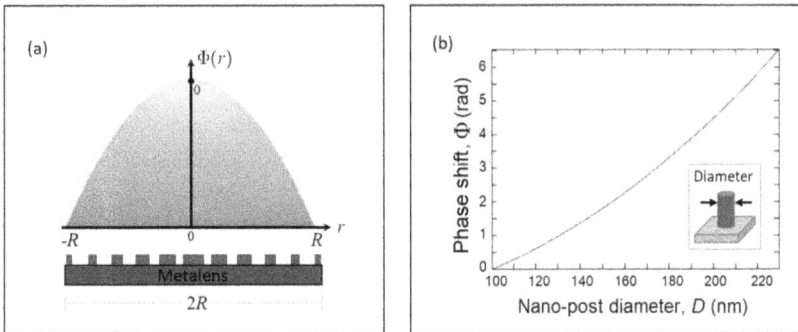

Figure 2.5. (a) Shows the phase profile in a metalens of radius R, which determines its focusing properties and the focal distance f. (b) Shows a simple relationship between the phase shift and the diameter of a nanopillar in a dielectric metasurface [2]. Figure reproduced from [5], copyright (2022) European Physical Society.

obtained by numerical calculations with finite difference time domain (FDTD) methods or rigorous coupled wave analysis (RCWA).

For example, the phase shift introduced by some dielectric nanocylinders or nanoposts may be approximated by $\Phi = K_1 D^2 + K_2 D + K_3$, where D is the post diameter, and K_1–K_3 are constants that may depend on several physical and geometrical parameters (such as meta-atom refractive index, meta-atom height, etc). Figure 2.5(b) shows a typical example of this relationship, where the phase shift Φ varies with the diameter D of nanopillars. Obtaining this relationship allows the reproduction of the desired phase profile $\Phi(r)$ by radially varying the diameter $D(r)$ of nanoposts across the metalens surface. For example, if the phase shift function is hypothetically $\Phi = K_1 D^2 + K_2$, a hyperbolic focusing metalens should have a diameter distribution given by:

$$D(r) = \sqrt{\frac{\Phi(r) - K_2}{K_1}} = \sqrt{\frac{k}{K_1}(f - \sqrt{r^2 + f^2}) - \frac{K_2}{K_1}}, \tag{2.13}$$

where r is the distance from the metalens center, K_1 is a constant dependent on the nanoelements, and k is the wave number. Generally, this $D(r)$ is the nanopillar diameter distribution function. In general this $D(r)$ is obtained by assuming that the phase shift is monotonic and by inverting the phase shift function. Note that the root of Eq. (2.13) becomes negative for all values of r; however, this is because we introduced Eq. (2.12) into Eq. (2.13) in a simplified manner for didactic purposes. To avoid this issue for practical purposes, the phase profile of Eq. (2.12) must be made positive by adding a constant (which has no effect on light focusing as explained in Sec. 2.2.3). Therefore, instead of using $\Phi(r)$ in Eq. (2.13), $\Phi(r) - \Phi(R)$ must be used, where R is the metalens radius.

The phase profile of a metalens typically includes phase values much larger than 2π, and the periodicity of electromagnetic waves with the phase must be considered. This periodicity is given by $\cos(\Phi) = \cos(\Phi \pm 2\pi n)$, where n is an integer. Therefore,

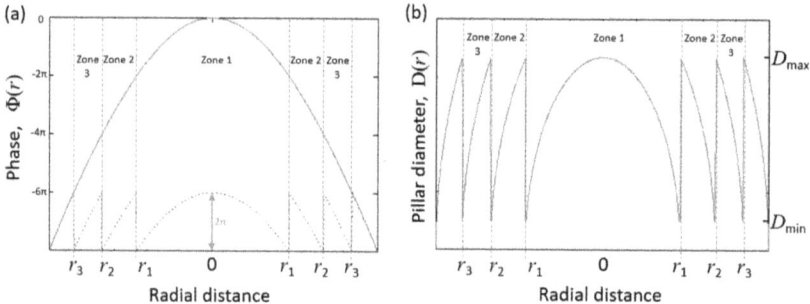

Figure 2.6. Phase profile of a metalens with phase values larger than 2π. (a) The phase profile of the metalens as a function of distance from the optical axis. The phase profile is divided into multiple zones, each with a phase interval of 2π. Each zone is characterized by a continuous phase interval of 2π, with phase discontinuities occurring at the zone boundaries. (b) An example of the nanopillar diameter distribution, $D(r)$, as a function of the radial distance, r, on the metalens. This diameter distribution is also divided into zones with periodic values.

the phase profile must be divided into multiple radial zones (figure 2.6(a)), each with a phase interval of 2π. Each radial zone is characterized by a continuous phase interval of 2π. This division into zones determines the meta-atom periodic distribution, where $D(\Phi) = D(\Phi \pm 2\pi n)$. For example, consider the nanopillar diameter distribution $D(r)$ given by equation (2.13), which is a function of the radial distance, r, on the metalens. The resulting diameter distribution is also divided into zones with periodic values, as shown in figure 2.6(b).

2.3 Relationship with the classical lens

Introducing the field of metalenses by a geometric approach may be complemented by connecting the phase profile with refractive optics. In this context it may be illustrative to compare a metalens with a classical lens, in particular with a thin lens. A refractive lens is usually composed of glass made with curved surfaces. Thin lens approximation states that its thickness along the optical axis is negligible. A lens is a thin lens if light entering at one point on the first face exits at approximately the same point on the second face. Thus a thin lens shifts the incident wavefront by an amount proportional to the effective thickness of the lens at each point. A fundamental result and concept of refractive optics in the paraxial approximation, is that a thin lens introduces a phase profile given by [11]:

$$\Phi(r) = \Phi_0 - \frac{k_0(n-1)r^2}{2}\left(\frac{1}{R_1} - \frac{1}{R_2}\right), \tag{2.14}$$

where R_1 and R_2 are the first and second radii of curvature of the refractive lens, k_0 is the vacuum wave number, Φ_0 is the phase shift introduced on the optical axis, and n is the refractive index of the lens.

We can see that equation (2.14) contains the well-known lens makers' equation:

$$\frac{1}{f} = (n-1)\left(\frac{1}{R_1} - \frac{1}{R_2}\right), \tag{2.15}$$

so the phase profile for a convergent refractive lens, in the paraxial and thin-lens approximations, can be written as:

$$\Phi(r) = \Phi_0 - \frac{k_0 r^2}{2f}, \tag{2.16}$$

Now consider the hyperbolic phase profile of a metalens, given by equation (2.12). In the paraxial approximation, the radial distance r is quite shorter than f, and we can approximate the term $(r^2 + f^2)^{0.5} \approx f + r^2/2f$. Therefore, the hyperbolic phase profile of a convergent metalens, in the paraxial approximation, can be written as:

$$\Phi(r) = -\frac{k_0 r^2}{2f}, \tag{2.16}$$

where $n_2 = 1$, and $k = k_0$. We note that the phase profiles of a refractive lens and a metalens are practically the same under the considered approximations. This is because the physical condition for bending light is the phase gradient $d\Phi/dr$, not the phase Φ itself, as explained in section 2.2.3. So, in the paraxial approximation, in order to focus light, both a metalens and a refractive thin lens have the same phase profile, or more precisely the same phase gradient $d\Phi/dr$. We could conclude that any lens, in addition to refractive lenses or metalenses, that imparts this phase transformation can focus light under the thin-lens and paraxial approximations [11]. In other words, the metalens phase profile, necessary for focusing and imaging, may be the same as that of refractive and diffractive lenses, but only in the thin-lens and paraxial approximations.

2.4 Meta-atoms

A metalens is usually composed of dielectric meta-atoms that exhibit no losses within the desired bandwidth and have a high refractive index. A high refractive index is essential for effectively confining light within the nanoelements and increasing the phase shift introduced, which enhances the focusing efficiency. Moreover, a high refractive index significantly simplifies the manufacturing process by reducing the necessary height of the meta-atoms needed to achieve a 2π phase delay.

A meta-atom introduces a phase shift to the light that is transmitted through it. There are three main mechanisms used to introduce the phase delay in dielectric metalenses through nanoelements: geometrical phase, propagation phase, and resonant phase. The fundamental principle of resonance tuning is Mie-type resonances. The principle behind the geometrical phase is the Pancharatnam–Berry phase, which is proportional to the orientation angles of asymmetric meta-atoms (nanofins) and requires circularly polarized light. The propagation phase works by accumulating phase through light propagation. These mechanisms are discussed in detail in chapter 4.

However, the finite size of nanoelements and the spacing between them cause the real phase profile of a metalens to differ slightly from the ideal (or target) phase profile. This discrepancy, along with considerations for aberration corrections, new application implementations, and other factors, has led to the development of numerous approaches and structures for metalenses reported in the literature. These are all aimed at designing and optimizing better metalenses.

2.5 Lens equation

A metalens has conjugate object and image points. Ideally, each object point being imaged maps perfectly to a point on the image. This is perfectly true for refractive lenses and metalenses, but only under the paraxial (small-angle) and thin-lens approximations. Let us derive the lens equation of a metalens, the equation that relates the object and image distances. Consider a point light source placed at point O in figure 2.7, where O is any point on the optical axis to the left of the metalens. Light rays that originate at O converge after transmitting from the metalens, and focus at the image point I, which is a real image of the object.

Consider a metalens surrounded by two transparent media with refractive indices n_1 and n_2. The object is in the medium with index of refraction n_1. Let us consider the paraxial rays leaving O. All such rays should be refracted at the metalens and be focused

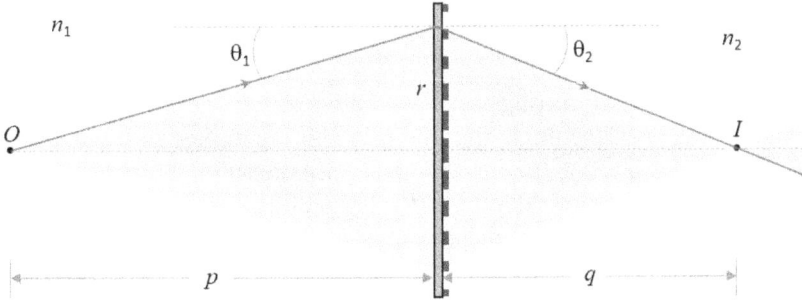

Figure 2.7. Metalens and conjugate points: image point I and object point O. Metalens focuses light rays through the image point I. A bundle of rays (in yellow) are leaving point O and are focusing to point I. Figure reproduced from [5], copyright (2022) European Physical Society.

at a single point I, the image point. Figure 2.7 shows a bundle of rays leaving point O and focusing to point I, and one ray (in red) is shown for the lens equation derivation.

Paraxial rays diverge from the object and make a small angle with the optical axis. Then, angles θ_1 and θ_2 are considered to be small in the generalized law of refraction, and the small-angle approximation $\sin\theta \approx \theta$ (in radians) can be used in equation (2.1), giving

$$\frac{1}{k_o}\frac{d\Phi}{dr} = n_2\theta_2 - n_1\theta_1, \tag{2.18}$$

In figure 2.7 one can see two right triangles that have a common vertical length r. In the small-angle approximation, $\tan\theta \approx \theta$, so we can approximate angles as follows:

$$\theta_1 \approx \tan\theta_1 = \frac{r}{p}, \tag{2.19}$$

$$\theta_2 \approx \tan\theta_2 = \frac{-r}{q}, \tag{2.20}$$

Equation (2.20) has a negative sign because the angle θ_2 is negative below the optical axis, a classical convention of signs. As explained in section 2.3, the hyperbolic phase profile of a convergent metalens, in the paraxial approximation, can be written as $\Phi(r) \approx -(kr^2)/2f$, so we can write the approximate phase gradient by:

$$\frac{1}{k_o}\frac{d\Phi}{dr} \approx -n_2\left(\frac{r}{f}\right), \tag{2.21}$$

If we substitute equations (2.19)–(2.21) into equation (2.18) and divide through by r, we obtain

$$\frac{n_1}{p} + \frac{n_2}{q} = \frac{n_2}{f}, \tag{2.22}$$

In most cases the object and image are in the same medium ($n_2 = n_1$), and equation (2.22) can be written in a form identical to the thin lens equation [3].

$$\frac{1}{p} + \frac{1}{q} = \frac{1}{f},$$ (2.23)

This equation shows that an object distance p is related with an image distance q, which is independent of the angle that the paraxial ray makes with the axis, i.e., all paraxial light focus at one point I. This approximation may well provide an adequate solution for practical purposes. The above derivation of the equation that relates the object and image distance in a metalens shows that any type of lens (thin enough, and in the paraxial approximation) should have the same lens equation for imaging.

2.6 Ray-tracing

In general, ray tracing methods through metasurfaces require the 3D direction of the transmitted/reflected beams through a metasurface with 2D phase gradient [6, 12]. In particular, the direction vectors of the reflected and refracted light is needed. These direction vectors may simplify the 3D ray tracing through systems of metalenses and other devices with metasurfaces. Let us derive the vector form equations for the direction of transmitted and reflected beams at a metasurface with arbitrary 2D phase spatial profiles, which are valid for any direction of incident light. These expressions must be functions of three known quantities: the unit vector normal to the metasurface, the direction vector of the incident beam, and the gradient of the phase profile.

By utilizing geometric characteristics and the definitions of the dot and cross product among any three unit vectors (k_1, k_2, n), it is possible to express the unit vectors of the incident ray and refracted ray as follows:

$$\begin{cases} k_1 = (n \cdot k_1)n + (n \times k_1) \times n \\ k_2 = (n \cdot k_2)n + (n \times k_2) \times n \end{cases}$$ (2.24)

where k_1 and k_2 are the unit vectors along the directions of the incident ray and refracted (or reflected) ray, and n is the unit vector along the normal to the metasurface (see figure 2.3). The next step is to substitute equation (2.24) in the 3D vector form of refraction–reflection law equation (2.11), and we obtain:

$$k_2 = (n \cdot k_2)n + \beta \nabla \Phi + \mu(n \times k_1) \times n,$$ (2.25)

here $\beta = 1/k_0 n_2$, and $\mu = n_1/n_2$, where n_1 is the refractive index of the incident medium and n_2 of the refractive index of transmitted/reflected medium.

Now note that the vector $(n \cdot k_2)n$ is normal to the metasurface plane, and then it is orthogonal to the vector $\beta \nabla \Phi + \mu(n \times k_1) \times n$, which lies on the metasurface. Therefore, by Pythagoras theorem, the norm of the unitary vector k_2 is:

$$1 = (n \cdot k_2)^2 + |\beta \nabla \Phi + \mu(n \times k_1) \times n|^2,$$ (2.26)

finally, by replacing the dot product of this equation in equation (2.25), the direction of the refracted/reflected ray in terms of the direction of incident ray, phase gradient, and the metasurface normal, is:

$$k_2 = \mu(\boldsymbol{n} \times \boldsymbol{k}_1) \times \boldsymbol{n} + \beta \, \nabla \, \Phi \pm \sqrt{1 - |\beta \, \nabla \, \Phi + \mu(\boldsymbol{n} \times \boldsymbol{k}_1) \times \boldsymbol{n}|^2} \, \boldsymbol{n}, \quad (2.27)$$

where the plus/minus sign denotes two possible values that depend if the root is positive or negative. The minus sign is for transmission, and the positive sign is for reflection. The sign is negative for the transmitted beam because it is propagating away from the backside of the metasurface, or because k_2 has the same direction along the normal as that of k_1, i.e. the sign of $\boldsymbol{n} \cdot \boldsymbol{k}_2$ is negative and equal to the sign of $\boldsymbol{n} \cdot \boldsymbol{k}_1$ (see figure 2.3). In case of reflection, the root in equation (2.27) is positive. This is because the reflected beam is propagating in the same side of the metasurface, or because the direction of the reflected beam has the opposite direction along the normal as that of the incident beam, i.e., the sign of $\boldsymbol{n} \cdot \boldsymbol{k}_2$ is positive and opposite to the sign of $\boldsymbol{n} \cdot \boldsymbol{k}_1$. In addition, reflection must consider $n_2 = n_1$.

Using the relation between dot and cross products, the direction of the refracted/reflected beam may be written as dot products:

$$k_2 = \mu[\boldsymbol{k}_1 - (\boldsymbol{n} \cdot \boldsymbol{k}_1)\boldsymbol{n}] + \beta \, \nabla \, \Phi \pm \sqrt{1 - |\beta \, \nabla \, \Phi + \mu(\boldsymbol{k}_1 - (\boldsymbol{n} \cdot \boldsymbol{k}_1)\boldsymbol{n})|^2} \, \boldsymbol{n}, \quad (2.28)$$

where the plus/minus sign denotes reflection and transmission, respectively.

Equations (2.27) and (2.28) have the ability to describe the transmitted or reflected beams of light through a metasurface. These equations agree with their scalar versions [6], giving the same results. Also, these equations reduce to the vector form of classical refraction in purely refractive interfaces by doing $\nabla \Phi = 0$ [13].

Problems

2.1 **Dual-Focus Metalens I.** Deduce a phase profile for a metalens with two focal lengths f_1 and f_2, and a diameter of D for the same wavelength λ.

2.2 **Dual-Focus Metalens II.** Consider a metalens with a hyperbolic phase profile (equation 2.12) that has two focal lengths f_1 and f_2 for two wavelengths λ_1 and λ_2, respectively. Demonstrate that the second focal length f_2 is approximately given by $f_2 \approx (\lambda_1/\lambda_2)f_1$ if the metalens diameter is notably shorter than both f_1 and f_2.

2.3 **Metamirror images.** Deduce, within the paraxial approximation, the lens equation for a focusing metamirror that relates an object point with its reflected image point.

2.4 **Off-Axis Metalens Focusing I.** Analyze the behavior of a metalens with a hyperbolic phase profile when the incident light is not normal to the lens surface but at an angle θ. Derive the effective focal length as a function of the incident angle. Determine the angular tolerance of the metalens, i.e. how the focal length varies with the incident angle of light.

2.5 **Off-Axis Metalens Focusing II.** Deduce the phase profile of a metalens for off-axis incidence at an angle θ.

2.6 **Generalized Snell's Law for Metasurfaces.** Consider a metasurface designed to refract incident light at a wavelength λ according to a prescribed phase gradient. Given an incident direction by angles (θ_i, φ_i) and a desired refracted angles (θ_t, φ_t), calculate the required phase gradient $\nabla \Phi$ on the metasurface.

Design a metasurface that redirects an incident beam at a wavelength. Specify the phase gradient and describe the arrangement of the meta-atoms.

Bibliography

[1] Khorasaninejad M and Capasso F 2017 Metalenses: versatile multifunctional photonic components *Science* **358** eaam8100

[2] Chen W T, Zhu A Y and Capasso F 2020 Flat optics with dispersion-engineered metasurfaces *Nat. Rev. Mater.* **5** 604–20

[3] Hecht E 2002 *Optics* (San Francisco, CA: Addison-Wesly)

[4] Yu N *et al* 2011 Light propagation with phase discontinuities: generalized laws of reflection and refraction *Science* **334** 333

[5] Moreno I 2022 Optics of the metalens *Eur. J. Phys.* **43** 065302

[6] Castañeda-Almanza C P and Moreno I 2022 Ray tracing in metasurfaces *Opt. Continuum* **1** 958–64

[7] Hu J, Xie J, Tian S, Guo H and Zhuang S 2020 Snell-like and Fresnel-like formulas of the dual-phase-gradient metasurface *Opt. Lett.* **45** 2251–4

[8] Aieta F, Genevet P, Yu N F, Kats M A, Gaburro Z and Capasso F 2012 Out-of-plane reflection and refraction of light by anisotropic optical antenna metasurfaces with phase discontinuities *Nano Lett.* **12** 1702–6

[9] Kim Y, Lee G-Y, Sung J, Jang J and Lee B 2022 Spiral metalens for phase contrast imaging *Adv. Funct. Mater.* **32** 2106050

[10] Bouillon C, Borne J, Ouellet-Oviedo E and Thibault S 2024 Semi-analytical models to engineer a metalens composed of various meta-atoms *J. Opt. Soc. Am.* B **41** 644–52

[11] Goodman J W 1996 *Introduction to Fourier Optics* 2nd edn (New York: McGraw-Hill)

[12] Gutiérrez C E, Pallucchini L and Stachura E 2017 General refraction problems with phase discontinuities on nonflat metasurfaces *J. Opt. Soc. Am.* A **34** 1160–72

[13] Tkaczyk E R 2012 Vectorial laws of refraction and reflection using the cross product and dot product *Opt. Lett.* **37** 972–4

IOP Publishing

Introduction to Metalens Optics

Ivan Moreno

Chapter 3

Simulation tools

Simulations play a crucial role in understanding and designing meta-atoms and metalenses in the realm of optics. The shape and size of individual nanoelements must be determined considering phase shift and light transmission. Light propagation through a nanoelement is governed by several optical effects simultaneously, and computing all of these effects is essential for metasurface design, which necessitates rigorous simulation methods such as finite-difference time-domain (FDTD) simulations. Analytical solutions for the transmitted or reflected light from arbitrary-shaped nanoelements do not exist, except for the simple case of spherical nanoparticles; therefore, numerical simulations are required to analyze and design metalenses. To accomplish these simulations, specialized computational tools are employed to model and analyze the behavior of these optical structures at the nanoscale and microscale levels (figure 3.1). In this chapter, we present some frequently used modeling methods of subwavelength structures. The transfer matrix method is a useful simulation tool in the analysis of layered dielectric media or multilayers, but its one-dimensional nature makes it not suitable for meta-atoms and metalenses. In the area metasurfaces, the prominent numerical simulation methods are FDTD, and rigorous coupled-wave analysis (RCWA). The FDTD method serves as powerful tool for simulating meta-atoms, metasurfaces, and metalenses. The RCWA method serves as a rapid simulation tool for optical characterizing meta-atoms and periodic metasurfaces. These numerical approaches enable researchers and designers to explore the optical behavior of nanostructured devices with precision and efficiency. There are several advanced simulation tools like Ansys Lumerical and MetaOptic Designer by Synopsys, which efficiently implement FDTD and RCWA methods for the optical development and optimization of metalenses.

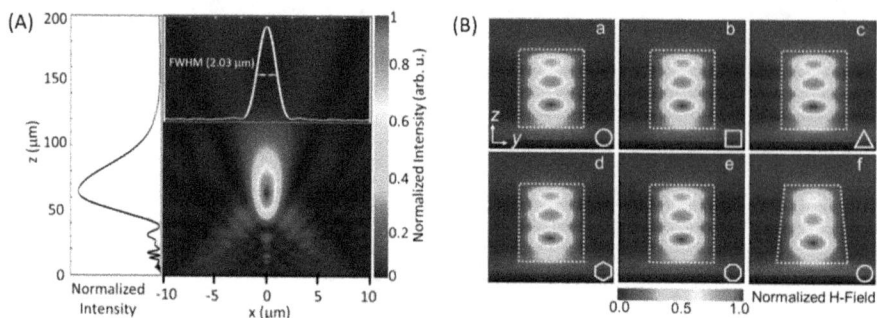

Figure 3.1. Numerical computions of electromagnetic fields propagating through (A) metalens, and (B) meta-atoms. (A) shows the intensity profile of the focal spot of a metalens in the xz-plane. The inset plots show the intensity in the z-axis (left), and the intensity across the focal spot in the x-axis (upper). Reprinted with permission from [2]. Copyright 2023 Optica Publishing Group. (B) illustrates the electromagnetic propagation through meta-atoms of different cross-section geometries (a–f). It is the normalized magnetic field (H) intensity in the propagation yz-plane. The white dashed lines represent the edges of the meta-atoms, and the yellow insets is the geometry. Reprinted with permission from [1]. Copyright (2024) American Chemical Society.

3.1 Finite-difference time-domain (FDTD)

The finite-difference time-domain (FDTD) method stands as a widely-used numerical technique for simulating the intricate behavior of complex optical structures such as meta-atoms, metasurfaces, and metalenses [3–5]. This method is a 3D full-wave electromagnetic solver. The FDTD method divides space into a three-dimensional grid and employs differential electromagnetic equations to compute the temporal and spatial evolution of electromagnetic fields. By discretizing Maxwell's equations, FDTD accurately models the interaction of light with photonic nanostructures.

First introduced during the 1960s by Kane S Yee, FDTD is an algorithmic approach to solving Maxwell's equations in the field of computational electromagnetics. Conceived in the 20th century, this method allowed for the analysis and design of complex photonic devices of the 21st century.

The FDTD method is a fully vectorial approach that naturally provides both time domain and frequency domain information, offering unique insights for analyzing electromagnetic and photonic problems. By operating in the time domain, transient data from a single simulation can be transformed into the frequency domain, enabling the extraction of wideband responses. Within FDTD, the simulation domain is defined as the space bounded by the simulation region and discretized by the mesh (figure 3.2). During an FDTD simulation, electromagnetic fields are computed from Maxwell's equations in each mesh cell, and solutions are iteratively advanced through time. Spatial discretization facilitates the representation of intricate geometries and structures, while temporal discretization captures the temporal evolution of electromagnetic fields.

FDTD solves the Faraday's and Ampere's laws [5], i.e., the two Maxwell's curl equations, because these equations rule electromagnetic field propagation. In other words, the Faraday's law rules how a changing magnetic flux induces a curly electric

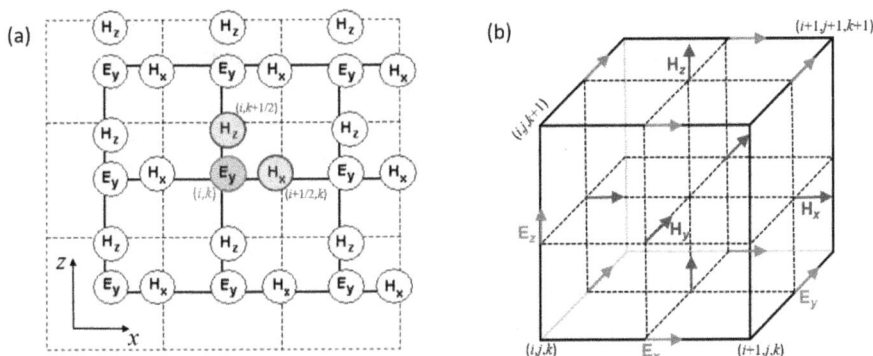

Figure 3.2. Schematic of Yee grid, showing the interleaving of the E and H fields for the: (a) 2D FDTD, and (b) 3D FDTD formulations. These fields are computed from Maxwell's curl equations in each mesh cell.

field, and the Ampere's law rules how a a changing electric flux induces a curly magnetic field. This electromagnetic field induction in time sequence is the key idea of the FDTD method, which solves the following Maxwell's equations:

$$\frac{\partial E}{\partial t} = \frac{1}{\varepsilon} \nabla \times H \tag{3.1}$$

$$\frac{\partial H}{\partial t} = -\frac{1}{\mu} \nabla \times E, \tag{3.2}$$

where E and H, are the electric and magnetic fields, respectively. Here ε is the permittivity, and μ is the permeability. E and H are vectors in three dimensions, so in general, equations (3.1) and (3.2) represent three equations each, having six electromagnetic field components in the three-dimensional case: E_x, E_y, E_z, H_x, H_y, and H_z. This chapter focuses on analyzing Maxwell's equations for non-magnetic dielectrics, where material conductivity $\sigma = 0$ and permeability μ is constant.

3.1.1 Two-dimensional FDTD approach

Let us now apply the finite-difference numerical approximation to the Maxwell's curl equations. For simplicity, let us explain the simpler two-dimensional case using only three components (E_y, H_x, H_z), given by

$$\frac{\partial E_y}{\partial t} = \frac{1}{\varepsilon} \left(\frac{\partial H_x}{\partial z} - \frac{\partial H_z}{\partial x} \right) \tag{3.3}$$

$$\frac{\partial H_x}{\partial t} = \frac{1}{\mu} \frac{\partial E_y}{\partial z}, \quad \frac{\partial H_z}{\partial t} = -\frac{1}{\mu} \frac{\partial E_y}{\partial x}, \tag{3.4}$$

Here, variations in electromagnetic properties occur within the xz-plane, with waves propagating in the z-direction. The material is infinite in the y-direction, allowing for the removal of all $\partial/\partial y$ derivatives from Maxwell's equations, thereby dividing them into two independent equation groups: TE and TM polarization. For this explanation, we focus on TE polarization ($E_x = 0$), which involves H_x, E_y, and H_z.

In FDTD, spatial and time derivatives in the two Maxwell's curl equations are replaced by central finite-difference approximations of derivatives, for example:

$$\frac{\partial E(x_o)}{\partial x} \cong \frac{E(x_o + \Delta x/2) - E(x_o - \Delta x/2)}{\Delta x}, \tag{3.5}$$

where the derivative is evaluated at the x_o point. Therefore, equations (3.3) and (3.4) are solved at different locations and time in the space of interest, where both time and space are discretized. It is important to have a systematic interleaving of the fields to be calculated. This is illustrated in 2D and 3D grids in figure 3.2. The 2D mesh grid has the fields $E_y(i,k)$, $H_x(i,k)$ and $H_z(i,k)$ at the (i,k)-th unit cell in the xz-plane. Rewriting equation (3.3) in the finite-difference scheme gives:

$$\frac{E_y^n(i, k) - E_y^{n-1}(i, k)}{\Delta t} = \frac{1}{\varepsilon(i, k)} \left[\frac{H_x^{n-1/2}(i, k + 1/2) - H_x^{n-1/2}(i, k - 1/2)}{\Delta z} - \frac{H_z^{n-1/2}(i + 1/2, k) - H_z^{n-1/2}(i - 1/2, k)}{\Delta x} \right], \tag{3.6}$$

where n superscript indicates the time steps, and $\varepsilon(i,k)$ is the position-dependent dielectric permittivity. Indices i and k are the lattice's positions along the x-, and z-directions. Finite difference Δt is the time evolution step, and Δx and Δz are the distance steps along the x-, and z-directions. Here the locations of the E_y fields are associated with integer values of i and k. The magnetic fields H_x and H_z are located at the unit cell edges, and associated with $(i,k \pm 0.5)$ and $(i \pm 0.5, k)$ indices, respectively.

From equation (3.6) the electric field E_y at the location $(i\Delta x, k\Delta z)$ and time $n\Delta t$ can be determined. Additionally, by applying the finite-difference scheme to equation (3.4), the associated magnetic fields can also be derived. These three electromagnetic fields are numerically given by:

$$E_y^n(i, k) = E_y^{n-1}(i, k) + \frac{\Delta t}{\varepsilon(i, k)} \left[\frac{H_x^{n-1/2}(i, k + 1/2) - H_x^{n-1/2}(i, k - 1/2)}{\Delta z} - \frac{H_z^{n-1/2}(i + 1/2, k) - H_z^{n-1/2}(i - 1/2, k)}{\Delta x} \right], \tag{3.7}$$

$$H_x^{n+1/2}(i, k + 1/2) = H_x^{n-1/2}(i, k + 1/2) + \frac{\Delta t}{\mu} \left[\frac{E_y^n(i, k + 1) - E_y^n(i, k)}{\Delta z} \right], \tag{3.8}$$

$$H_z^{n+1/2}(i + 1/2, k) = H_z^{n-1/2}(i + 1/2, k) - \frac{\Delta t}{\mu} \left[\frac{E_y^n(i + 1, k) - E_y^n(i, k)}{\Delta x} \right], \tag{3.9}$$

These field components are computed at slightly offset positions within the grid cell, resulting in an interleaved calculation of the electromagnetic fields in both space and time. In essence, the new value of E_y is derived from the preceding value of E_y, as well as from the most recent values of H_x and H_z. This methodology resumes the 2D FDTD approach for the TE polarization case. As mentioned, it employs central difference numerical approximations for spatial and temporal derivatives, with spatial sampling conducted at subwavelength scales.

3.1.2 Three-dimensional FDTD approach

The three-dimensional FDTD simulation is very much like two-dimensional simulation, only all six electromagnetic fields instead of three must be considered. The 3D FDTD uses 3D unit cells instead of 2D unit cells, and other logistical issues must be considered. In general, metalenses and meta-atoms require 3D FDTD simulations because the computational space is volumetric.

In 3D FDTD simulations, the unit cell is a cubic box, where the spatial steps are Δx, Δy, and Δz. As illustrated in figure 3.2(b), the positions of the E and H field components within the cubic unit cell of the mesh space lattice are depicted. The distribution of field components to be computed in space is as follows: E_x $(i + 1/2, j, k)$, E_y $(i, j + 1/2, k)$, E_z $(i, j, k + 1/2)$, H_x $(i, j + 1/2, k + 1/2)$, H_y $(i + 1/2, j, k + 1/2)$, H_z $(i + 1/2, j + 1/2, k)$. This spatial distribution indicates that the E and H components are interleaved at intervals of $0.5\Delta x$, $0.5\Delta y$, and $0.5\Delta z$ in space to facilitate algorithm implementation.

Using all the vector components of the Maxwell's equations (equations (3.1) and (3.2)) and the finite-difference approximations for the various partial derivatives, the resulting scalar partial differential equations yield the following solutions. The electric field components are given by

$$E_x^{n+1}\left(i + \frac{1}{2}, j, k\right) = E_x^{n}\left(i + \frac{1}{2}, j, k\right) + \frac{\Delta t}{\varepsilon\left(i + \frac{1}{2}, j, k\right)}$$

$$\left[\frac{H_z^{n+1/2}\left(i + \frac{1}{2}, j + \frac{1}{2}, k\right) - H_z^{n+1/2}\left(i + \frac{1}{2}, j - \frac{1}{2}, k\right)}{\Delta y} - \frac{H_y^{n+1/2}\left(i + \frac{1}{2}, j, k + \frac{1}{2}\right) - H_y^{n+1/2}\left(i + \frac{1}{2}, j, k - \frac{1}{2}\right)}{\Delta z}\right], \quad (3.10)$$

$$E_y^{n+1}\left(i, j + \frac{1}{2}, k\right) = E_x^{n}\left(i, j + \frac{1}{2}, k\right) + \frac{\Delta t}{\varepsilon\left(i, j + \frac{1}{2}, k\right)}$$

$$\left[\frac{H_x^{n+1/2}\left(i, j + \frac{1}{2}, k + \frac{1}{2}\right) - H_x^{n+1/2}\left(i, j + \frac{1}{2}, k - \frac{1}{2}\right)}{\Delta z} - \frac{H_z^{n+1/2}\left(i + \frac{1}{2}, j + \frac{1}{2}, k\right) - H_z^{n+1/2}\left(i - \frac{1}{2}, j + \frac{1}{2}, k\right)}{\Delta x}\right], \quad (3.11)$$

$$E_z^{n+1}\left(i, j, k + \frac{1}{2}\right) = E_z^{n}\left(i, j, k + \frac{1}{2}\right) + \frac{\Delta t}{\varepsilon\left(i, j, k + \frac{1}{2}\right)}$$

$$\left[\frac{H_y^{n+1/2}\left(i + \frac{1}{2}, j, k + \frac{1}{2}\right) - H_y^{n+1/2}\left(i - \frac{1}{2}, j, k + \frac{1}{2}\right)}{\Delta x} - \frac{H_x^{n+1/2}\left(i, j + \frac{1}{2}, k + \frac{1}{2}\right) - H_x^{n+1/2}\left(i, j - \frac{1}{2}, k + \frac{1}{2}\right)}{\Delta y}\right]. \quad (3.12)$$

All H field quantities on the right-hand side are evaluated at time-step $n + 1/2$, and E values at time-step n. Similarly, the finite-difference expressions for the magnetic field components located about the unit cell are:

$$H_x^{n+1/2}\left(i, j + \frac{1}{2}, k + \frac{1}{2}\right) = H_x^{n-1/2}\left(i, j + \frac{1}{2}, k + \frac{1}{2}\right) + \frac{\Delta t}{\varepsilon\left(i, j + \frac{1}{2}, k + \frac{1}{2}\right)}$$

$$\left[\frac{E_y^n\left(i, j + \frac{1}{2}, k + 1\right) - E_y^n\left(i, j + \frac{1}{2}, k\right)}{\Delta z} - \frac{E_z^n\left(i, j + 1, k + \frac{1}{2}\right) - E_z^n\left(i, j, k + \frac{1}{2}\right)}{\Delta y}\right], \quad (3.13)$$

$$H_y^{n+1/2}\left(i + \frac{1}{2}, j, k + \frac{1}{2}\right) = H_y^{n-1/2}\left(i + \frac{1}{2}, j, k + \frac{1}{2}\right) + \frac{\Delta t}{\varepsilon\left(i + \frac{1}{2}, j, k + \frac{1}{2}\right)}$$

$$\left[\frac{E_z^n\left(i + 1, j, k + \frac{1}{2}\right) - E_z^n\left(i, j, k + \frac{1}{2}\right)}{\Delta x} - \frac{E_x^n\left(i + \frac{1}{2}, j, k + 1\right) - E_x^n\left(i + \frac{1}{2}, j, k\right)}{\Delta z}\right], \quad (3.14)$$

$$H_z^{n+1/2}\left(i + \frac{1}{2}, j + \frac{1}{2}, k\right) = H_y^{n-1/2}\left(i + \frac{1}{2}, j + \frac{1}{2}, k\right) + \frac{\Delta t}{\varepsilon\left(i + \frac{1}{2}, j + \frac{1}{2}, k\right)}$$

$$\left[\frac{E_x^n\left(i + \frac{1}{2}, j + 1, k\right) - E_x^n\left(i + \frac{1}{2}, j, k\right)}{\Delta y} - \frac{E_y^n\left(i + 1, j + \frac{1}{2}, k\right) - E_y^n\left(i, j + \frac{1}{2}, k\right)}{\Delta x}\right], \quad (3.15)$$

All E field quantities on the right-hand side are evaluated at time-step n, and H values at time-step $n-1/2$. With these six iterative equations, the new value of an E (or H) field at any lattice point and time depends only on its previous value, and on the previous values of the H (or E) fields at adjacent points. Therefore, at any given time step, the computation of an electromagnetic field can iteratively computed.

3.1.3 Space and time step conditions

The choice of the space ($\Delta x, \Delta y, \Delta z$) and time ($\Delta t$) steps can affect the propagation characteristics of numerical waves in the FDTD method and then the numerical error. Therefore, both the space step and the time step must be bounded to ensure simulation accuracy and numerical stability. In general, to keep the accuracy high, the mesh must contain a minimum of 10 cells per wavelength, i.e., the side of unit cell should be $1/10\lambda$ or less. However, because some portions of the computational space are filled with materials of a certain refractive index, the wavelength in the material must be considered in the maximum space step (mesh size), given by

$$(\Delta x, \Delta y, \Delta z) \leqslant \frac{\lambda}{10n}, \quad (3.16)$$

where n is the maximum refractive index, and λ is the minimum wavelength in the simulated optical system. Reducing the spatial step decreases simulation error but also increases simulation time.

Additionally, when selecting spatial steps (Δx, Δy, Δz), it's crucial to consider the minimum size characteristics of meta-atoms on the metalens. Some metasurfaces or metalenses may incorporate very small meta-atoms with diameters or heights around $\lambda/10$. In such cases, the maximum spatial step (mesh size) should be smaller than the size of the smallest meta-atoms used.

Once the cell size is established, the time step also requires a minimum value to have numerical stability in the FDTD simulation. This is given by a criteria known as the Courant condition [5, 6]. In the case of cubic cells, where $\Delta x = \Delta y = \Delta z = \Delta$, the Courant condition is given by

$$\Delta t \geqslant \frac{\Delta}{v\sqrt{3}}, \tag{3.17}$$

where v is the speed of light in the medium. This is a fundamental stability limit, which indicates that the time step used in the FDTD simulation must be long enough to allow the electromagnetic waves to propagate across a unit cell at the speed of light. This limits how fast the computation can be. An electromagnetic wave cannot go faster than c/n in the propagating medium. To propagate a distance in a one-dimensional cell, light requires a minimum time of $\Delta t = \Delta x/v$. Within a two-dimensional cell, the time step must allow the propagation through the diagonal direction, which requires a time of $\Delta t = \Delta/(v\sqrt{2})$. In the three-dimensional case it requires $\Delta t = \Delta/(v\sqrt{3})$. For non-cubic cells, the Courant condition is:

$$\Delta t \geqslant \frac{1}{v\sqrt{\frac{1}{\Delta x^2} + \frac{1}{\Delta y^2} + \frac{1}{\Delta z^2}}}, \tag{3.18}$$

The Courant condition limits the simulation time step to be larger than the maximum length dimension of the unit cell divided by the speed of light.

In practical simulations, it is necessary to truncate the mesh space in certain finite regions so that they can be computed. This leads to the utilization of the concept of a perfectly matched layer (PML), an artificially engineered region composed of lossy media capable of absorbing waves at any angle of incidence, regardless of frequency and polarization. Additionally, other essential boundary conditions are employed in the simulation of metasurfaces. For instance, periodic boundary conditions prove useful when the electromagnetic field demonstrates clear periodicity, enabling the deduction of the field distribution of the entire space from one unit cell. The correct use of boundary conditions shortens simulation time and enhances simulation accuracy.

3.2 Rigorous coupled-wave analysis (RCWA)

RCWA, also known as the Fourier modal method (FMM), serves as a simulation tool for analyzing and designing meta-atoms, offering computational efficiency superior to FDTD for simulating the optical properties of nanoelements, albeit with less precision. It operates as a semi-analytical method to solve Maxwell's equations within periodic nanostructures, making it useful for simulating light propagation through 3D nanostructures with 2D periodicity [3, 9]. Based on Fourier diffraction theory, RCWA efficiently computes light scattering in periodic structures by solving coupled-wave equations. Compared to FDTD, RCWA typically yields shorter simulation times [7], primarily utilized for meta-atom simulations. RCWA is particularly effective in generating meta-atom libraries for metalens design, enabling calculation of phase shift and transmittance by varying meta-atom parameters such as nanorod diameter and height.

In contrast to the FDTD method, RCWA discretizes Maxwell's equations in Fourier space and numerically solves them by computing the full scattering matrix [9, 10]. While FDTD methods discretize the spatial domain in all three dimensions, RCWA only discretizes it in the x- and y-dimensions, excluding the propagation direction or z-dimension (figure 3.3). Since the cross-sectional shape of most meta-atoms remains constant along the propagation direction, z-direction discretization is unnecessary. Consequently, light propagation in the nanoelement can be viewed as a superposition of individual propagating modes experiencing phase shifts.

In RCWA, the nanoelement is divided into layers, and the light propagation modes of each layer are calculated within Fourier space [8, 9]. Electromagnetic fields within each layer are expanded into eigenmodes with simple exponential dependence ($\exp[iqz]$) in the propagation direction. These fields are represented using a Fourier basis in the interface plane, with each Fourier component coupled via the dielectric spatial distribution within a layer. The modes are then expanded into the modes of adjacent layers, determining light coupling between them. The modal expansion coefficients of each layer are related at every layer interface to satisfy field continuity conditions in the Fourier basis. This process allows propagation of a plane wave

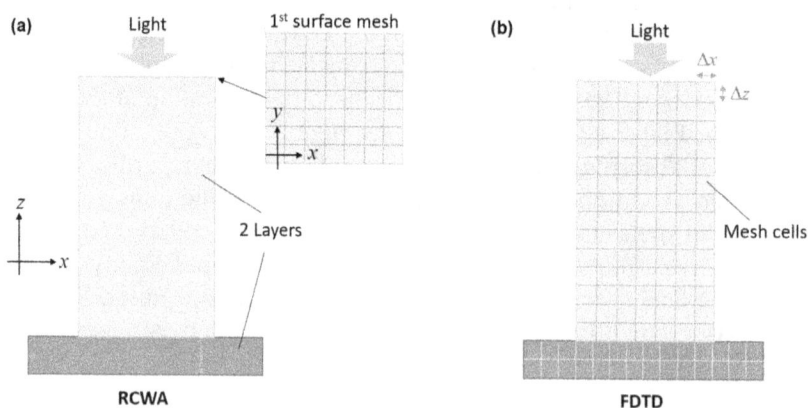

Figure 3.3. The discretization of a nanorod with a square cross-section into structured meshes. It illustrates the rod in yellow and the substrate in blue. (a) In RCWA, this nanoelement is divided into two layers. In the x and y-directions, the field is expanded into Bloch waves within each layer. In the z-direction, the solutions (electromagnetic field modes) within each layer are analytically propagated at a very low computational cost. (b) In FDTD, this nanoelement is decomposed into a structured or regular grid enabling a higher resolution of domain boundaries, but at a high computational cost. For simplicity, we omitted the mesh of the air within the unit cell that surrounds the nanorod.

with a specific frequency, direction, and polarization through the structure, achieved by expanding it into the modes within each layer.

The complete formulation of RCWA [3, 9, 10], along with the mathematical procedures and expressions to obtain the output field **E**, involve multiple Fourier transforms and matrices, making its mathematical explanation challenging for this book's purposes. However, let's outline some details of the RCWA formulation in a simplified calculation process:

1. The nanostructure is divided into layers, ensuring each layer is homogeneous in the propagation direction (usually the z-axis). If the nanostructure's cross-section shape varies through the propagation direction, the required number of layers increases, albeit at the cost of increased simulation time.

2. For each layer, electrical permittivity and electromagnetic field components are represented by Fourier series expansion. For example, the electric field expansion is:

$$E(r, z) = \sum_{G} \mathbf{S_G}(z)e^{i(\mathbf{k}+\mathbf{G})\cdot r}, \qquad (3.19)$$

where vector r represents the transverse coordinates in the xy-plane. **k** is the in-plane component of the k-vector of the incident plane wave. And **G** is a 2D reciprocal lattice vector of the Fourier domain. Fourier coefficients $\mathbf{S_G}$ determine the electric field, and usually those for the magnetic field are represented by $\mathbf{U_G}$.

3. Fourier modes in layers are calculated. This step starts by taking the spatial Fourier transform of Maxwell's equations in the xy-plane.

$$\begin{aligned}\mathcal{F}\{\nabla \times \mathbf{H} = -i\omega\varepsilon(r)\mathbf{E}\} \\ \mathcal{F}\{\nabla \times \mathbf{E} = i\omega\mathbf{H}\}\end{aligned} \qquad (3.20)$$

where permittivity $\varepsilon(r)$ can be represented in a Fourier series expansion. From equation (3.20) a series of scalar equations of Fourier coefficients $\mathbf{S_G}$ and $\mathbf{U_G}$ are obtained. Which solutions in matrix form give the S and U matrix for each layer.

4. Boundary conditions are used for neighbouring layers to form scattering matrices, i.e. S and U matrices are obtained for matching the boundary conditions between the layers.

5. Calculation of coupling coefficient for each Fourier component, i.e., computation of the field solution in the spatial frequency domain (k domain) for each layer.

6. Waves are propagated through the nanostructure by expanding modes into modes of neighboring layers. The solution for each layer is then propagated bi-directionally to calculate the S-matrix of the entire nanostructure, obtaining the total scattering matrix. This step allows for obtaining electromagnetic fields inside and outside the nanostructure, transmitted and reflected intensity, and their relative phases.

After calculating the scattering matrix, incident light can be propagated through the nanostructure. Due to the nanostructure's geometry, an incident plane wave diffracts into a finite set of plane waves referred to as 'grating orders.' With the S-matrix, the transmitted and reflected incident power fractions, power in each diffraction order, and electric and magnetic fields inside and outside the nano-structure can be calculated.

RCWA is a potent, rapid semi-analytical method for numerically investigating the optical properties of periodic nanostructures in the infrared and visible regions. However, for more complex nanostructures or meta-atoms, issues such as numerical stability in matching boundary conditions between layers and convergence problems due to computation of products in truncated Fourier space must be considered [10, 11]. Correct Fourier factorization rules over the entire period for arbitrary shape nanostructures can enhance result convergence in these cases.

3.3 Simulation softwares

Advanced numerical simulations of meta-atoms, metasurfaces, and metalenses are essential for their efficient optical design, optimization, and fabrication. The history of optics has seen great changes, transitioning from early lens designs that were realized without computers to the modern metalens design era that relies on computational software, which is indispensable for feasible research. Perhaps the primary challenge of computational simulations for metasurfaces and metalenses is the computational time required. Therefore, one of the most common considerations when selecting a simulation software is its speed, i.e., how long it takes to complete the simulation and analysis. Consequently, software tools are crucial for metalens optical analysis because they offer reliable and time-saving techniques. Given the numerous challenges in simulation, the optical analysis of metalenses demands thorough analysis, necessitating efficient software tools and models for their rapid and effective optical design, analysis, and optimization.

The metalens nanoscale is explored using computers and photonics simulation software, enabling visualization, computation, and assessment of light propagation through the metalens. Most novel metalens discoveries and designs use advanced simulation softwares, which are fundamental tools in the creative process of finding new concepts, improvements and applications of flat optics, be it verfying the feasibility of great new ideas, simulating a proof-of-concept, optimizing a new design, or perhaps trying to verify a new metalens application. In all these cases a nano-photonic simulation software is necessary. Several software packages have been developed to simulate light propagation through meta-atoms and metalenses, assess their optical performance, simplify the optical design process, and maximize their efficiency. Below, we provide a brief overview to help identify suitable software tools.

PlanOpSim: PlanOpSim software enables the approximate calculation of large metalenses and metasurfaces [12, 13]. It is a powerful software for the design and simulation of metasurfaces, which allows for a larger area, accurate design in an interface that is easy to grasp, and faster in most cases than FDTD. PlanOpSim is characterized by an easy-to-use metalens software tool, that has the following features: (a) Metacell, an optical design of nanostructures using a full solution of Maxwell's equations. (b) Meta-Component, a unit for designing metasurfaces, metalenses, components and holograms using optimization methods. (c) Integrates metasurface components into optical systems with 3rd party ray-tracing and CAD softwares.

MAXIM: MAXIM is a freeware tool that utilizes rigorous coupled-wave analysis (RCWA). This tool is an attractive option, featuring an easy-to-understand user interface and lightweight computation on personal computers [14]. It has an intuitive graphical user interface that improves the accessibility of the software to who are not familiar with computer programming. Its computation performance was evaluated for several didactic examples of dielectric metasurfaces, which are the main application of MAXIM. Nevertheless, a comparison with commercial software based on the FDTD method shows that the computation results coincide closely with each other within 1% difference. Therefore, their features include easy accessibility, wide availability and high reliability. Its application is mainly on meta-atoms.

SMD (Simple Metalens Design) Tool: SMD Tool is a free optics software designed to assist engineers in building and simulating metalenses. Similar to MAXIM, it offers an intuitive user interface and lightweight computation on personal computers [15]. The tool consists of three modules: Target Phase Generator (Module 1), Unitcell Arranger (Module 2), and Lens Simulator (Module 3). Its main interface displays module buttons and a visual description of metalens parameters and coordinate systems. Users can set the size, focal length, and focal position of the desired metalens on Module 1. Module 3 performs computations based on Huygens' principle (Fresnel–Kirchhoff diffraction formula).

Meep (MIT Electromagnetic equation Propagation) Tool: Meep is a free and open-source electromagnetic software. It uses the FDTD method with perfectly matched layer or periodic boundary conditions for field computation [16, 17]. Because it is open-source, the tool has been expanded to include topology optimization and inverse design [18]. Meep software is a very popular tool among the optics and photonics communities, including the analysis and design of metalenses and metasurfaces. This software is based on finite-difference frequency-domain (FDFD) methods. Meep includes a materials library that contains predefined broadband, complex refractive indices.

COMSOL Multiphysics: COMSOL Multiphysics is a general-purpose simulation software with fully coupled multiphysics and single-physics modeling capabilities [19, 20]. It allows the incorporation of multiphysics processes to model metalenses, such as thermally configurable metalenses. In general, this software tool can incorporate other physical effects, for instance, stress-optical, electro-optical, and acousto-optical effects as well as electromagnetic heating. COMSOL Multiphysics incorporates advanced numerical methods in electromagnetics problems, such as the finite element method (FEM), method of moments (MoM), and ray tracing. This software gets very precise results, but it is not optimized for fast metalens calculations.

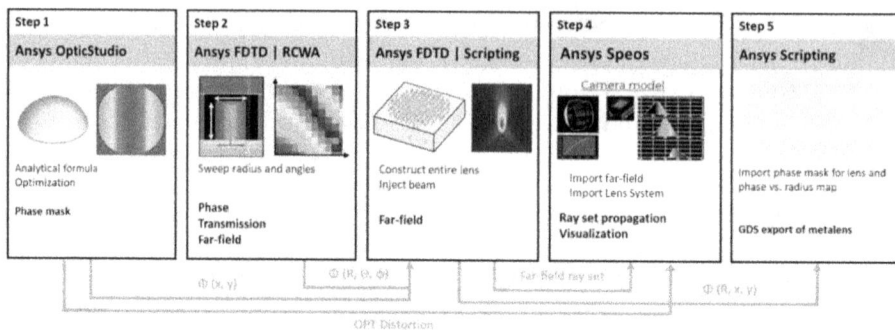

Step 1	Step 2	Step 3	Step 4	Step 5
Ansys OpticStudio	Ansys FDTD | RCWA	Ansys FDTD | Scripting	Ansys Speos	Ansys Scripting
Analytical formula Optimization Phase mask	Sweep radius and angles Phase Transmission Far-field	Construct entire lens Inject beam Far-field	Import far-field Import Lens System Ray set propagation Visualization	Import phase mask for lens and phase vs. radius map GDS export of metalens

Figure 3.4. Metalens system design workflow in a simulation software. It shows the different steps followed to design a metalens for a fiber-optic endoscope in Ansys Lumerical software. The design steps of this example are: (1) target phase profile calculation; (2) using the RCWA method to simulate meta-atoms, and get a database library of the amplitude, phase, and polarization for each meta-atom parameter; (3) using the target phase profile to choose the parameters that provide the best phase match, and assembling the desired metalens; (4) to use the ideal phase profile and simulated fiber-optic illumination in the physics-based ray tracing software for image visualization in a blood vessel; and (5) the metalens layout is saved to GDSII format for nanofabrication. Image reproduced from reference [22].

Ansys Lumerical: Ansys Lumerical software stands as one of the premier programs implementing the FDTD method for simulating photonics devices [21]. With Lumerical, users can design and analyze a broad spectrum of nanostructures, ranging from individual meta-atoms to complete metalenses. The software provides advanced tools for visualizing electromagnetic fields, dispersion analysis, design optimization, and performance evaluation. This tool allows one to design metalens systems (figure 3.4), for example a metalens based double-path fiber-optic endoscope [22]. The required phase shifts of meta-atoms are simulated with a RCWA solver by sweeps over geometric and physical parameters. Light propagation through the metalenses is simulated by FDTD. For large metalenses, it offers ray-tracing based approaches for system level analyses, and indirect field reconstruction techniques instead of direct FDTD simulations. In addition, the software offers image visualization of the simulated optical system in a 3D environment. Ansys Lumerical software incorporates advanced numerical methods such as FDTD, FEM, MoM, and RCWA.

Synopsys Tools: MetaOptic Designer, RSoft, and CODE V are Synopsys tools used for metaoptics design and simulation. MetaOptic Designer is the first fully automated commercial tool that provides inverse metalens design capabilities [23, 24]. This tool is fast, and it requires a few basic inputs, as the desired target light pattern to produce an optimized design. Their simulations are validated by FDTD, and has a built-in intelligence to make metalens design faster and easier. The optimization algorithm employs the adjoint method, which can practically handle millions of design variables. Light propagation is done with an efficient Fourier transform method (angular spectrum method). The phase shift of each meta-atom is stored in a database built with the FDTD method or RCWA using other Synopsys products. The RSoft tool allows RCWA and FDTD simulations of meta-atoms and metalenses. However, more efficient approximate algorithms, such as the beam propagation method (BPM) or beam synthesis propagation (BSP) through CODE V, are used to simulate the propagation through the metalens. With these software tools, users can model light propagation through meta-atoms and metalenses and analyze their optical properties such as reflectance, transmittance, dispersion, and phase.

MetaOptics: MetaOptics is an open-source software built using Python for rendering generic metasurface GDSII layouts [25]. This is not an electromagnetic simulation tool as the other commented softwares. The metalens layout is saved in GDSII, a data format typically used in nanofabrication methods. MetaOptics uses the FDTD simulated phase response meta-atom library to convert the desired phase profile into a metasurface GDSII layout for manufacturing. MetaOptics has a library of in-built FDTD data for most commonly used wavelengths in the visible and infrared spectrum. It also has an option to upload user-specific dimension versus transmission phase data for any choice of wavelength.

Simulations performed for metalens analysis and design are constrained to smaller lens diameters than those fabricated, due to computational resource limits. However, recent metalenses with diameters on the order of centimeters have been demonstrated. But the larger the metalens, the more computationally expensive accurate simulations become. Full wave numerical methods to solve Maxwell's equations, such as FDTD and FEM, require enormous computational resources. Therefore, several strategies to accelerate simulations have been proposed. For example, accurate simulation methods of metalenses with dimensions as large as 3000 times the operating wavelength on a personal computer [26]. Or for instance, ray-based metalens analysis tools [27], which bridge the gap between the metalens (milimeter-scale) and meta-atoms (nanometer-scale), and reduce the complexity of the metalens design. In addition, semi-analytical models to rapidly obtain the propagation characteristics of meta-atoms have been developed [28].

3.4 Problems

3.1 **Optimizing Meta-Atom Design with a FDTD Simulation.** Design and optimize a dielectric nanopillar for a metalens using a finite-difference time-domain (FDTD) simulation. Create a design for a meta-atom that provides a phase shift of π at a wavelength of 550 nm. Specify the geometric and material properties. Define the space and time step conditions to ensure accurate results without excessive computational load. Discuss the impact of different parameters on the phase shift and transmission.

3.2 **RCWA for Multi-Layer Meta-atom Design.** Use rigorous coupled-wave analysis (RCWA) to design a multi-layer dielectric nanorod that operates efficiently at a wavelength of 500 nm. Propose an initial design for the multi-layer structure, specifying the thickness and material of each layer. Set up an RCWA simulation to analyze the transmission and phase response of the meta-atom. Optimize the layer thicknesses and material properties to maximize transmittance and minimize chromatic dispersion. Compare the RCWA results with FDTD simulation results.

3.3 **Simulation Software Comparative Study.** Conduct a comparative study of different simulation software tools (e.g., Lumerical, COMSOL Multiphysics, and a free access software) for modeling a metalens or meta-atom. Simulate the same metalens or meta-atom at a wavelength of 633 nm using each simulation tool. Specify the geometric and material parameters. Ensure consistent boundary conditions and meshing across all tools. Compare the results in terms of focusing efficiency, computational time, and ease of use. Provide a detailed analysis of the strengths and weaknesses of each tool for metalens design.

3.4 **FDTD Simulation and Space-Time Step Conditions.** Perform a FDTD simulation of a metalens and analyze the impact of different space and time step conditions on the simulation accuracy and computational time. Vary the space and time step conditions to investigate their effects on the accuracy and stability of the simulation. Consider resolutions of λ, $\lambda/10$, $\lambda/20$, and $\lambda/40$ for spatial steps and corresponding time steps. Evaluate the simulation accuracy by comparing the focal spot size and field intensity for different step conditions.

Bibliography

[1] Basilio-Ortiz J C and Moreno I 2024 Unveiling invariant optical properties of dielectric meta-atoms *Nano Lett.* **24** 4987–92

[2] Basilio-Ortiz J C and Moreno I 2023 All dielectric reflective metalens based on multilayer meta-atoms *Opt. Lett.* **48** 5647–50

[3] Luo X 2019 Numerical modeling and intelligent designs *Engineering Optics 2.0* (Singapore: Springer)

[4] Vafaei R 2024 What is finite-difference time-domain (FDTD)? ANSYS Blog http://www.ansys.com/en-gb/blog/what-is-fdtd

[5] Sullivan D M 2000 *Electromagnetic Simulation using the FDTD Method* (New York: IEEE)

[6] Taflove A 1995 *Computational Electrodynamics: The Finite-Difference Time-Domain Method* 3rd edn (Boston, MA: Artech House)

[7] Jang H T and Yi J C 2021 Comparison of rigorous coupled-wave analysis and finite difference time domain method on dielectric gratings *IEEE Region 10 Symp. (TENSYMP) (Jeju, Republic of Korea, 2021)* pp 1–4

[8] Herbert K 2023 How RCWA solvers support ongoing innovation in optical technologies *ANSYS Blog.* https://www.ansys.com/blog/rcwa-supports-ongoing-optical-technology

[9] Liu V and Fan S 2012 S4: a free electromagnetic solver for layered periodic structures *Comput. Phys. Commun.* **183** 2233–44

[10] Cherqvist L 2023 Halldner, global optimization of optical metasurfaces using the RCWA method *Master Thesis* Chalmers University of Technology, Gothenburg, Sweden

[11] Zhu Z and Zheng C 2022 VarRCWA: an adaptive high-order rigorous coupled wave analysis method *ACS Photonics* **9** 3310–7

[12] Robben B and Penninck L 2024 Simulation methods for large-area meta-surfaces: comparison local periodic, overlapping domains, and full wave calculations *Proc. SPIE 12897, High Contrast Metastructures XIII* 1289709

[13] https://planopsim.com/

[14] Yoon G and Rho J 2021 MAXIM: metasurfaces-oriented electromagnetic wave simulation software with intuitive graphical user interfaces *Comput. Phys. Commun.* **264** 107846

[15] https://simplemetalens.com/

[16] Oskooi A F *et al* 2010 Meep: a flexible free-software package for electromagnetic simulations by the FDTD method *Comput. Phys. Commun.* **181** 687–702

[17] https://meep.readthedocs.io/en/latest/

[18] Hammond A M, Oskooi A, Chen M, Lin Z, Johnson S G and Ralph S E 2022 High-performance hybrid time/frequency-domain topology optimization for large-scale photonics inverse design *Opt. Express* **30** 4467–91

[19] https://comsol.com/

[20] Paasonen V 2023 Building a metalens design app with COMSOL multiphysics *COMSOL Blog.* https://comsol.com/blogs/building-a-metalens-design-app-with-comsol-multiphysics/

[21] https://ansys.com/

[22] Timar-Fulep C, Meyer C, Kim S-S and Thoene S 2024 Metalens design for fiber-optic endoscope *Proc. SPIE 12817, Advanced Photonics in Urology 2024* 128170D

[23] Xu C 2023 MetaOptic designer: automated design of metalenses *White Paper*

[24] https://synopsys.com/photonic-solutions/rsoft-photonic-device-tools.html

[25] Dharmavarapu R, Hock Ng S, Eftekhari F, Juodkazis S and Bhattacharya S 2020 MetaOptics: opensource software for designing metasurface optical element GDSII layouts *Opt. Express* **28** 3505–16

[26] Martins A, da Mota A F, Stanford C, Contreras T, Martin-Albo J, Kish A, Escobar C O, Para A and Guenette R 2024 Simple strategy for the simulation of axially symmetric large-area metasurfaces *J. Opt. Soc. Am.* B **41** 1261–9

[27] Ding Y and Stone B D 2023 Validation of a ray-based tool for metalens design and analysis *Proc. SPIE 12798, Int. Optical Design Conf. 2023* p 1279807

[28] Bouillon C, Borne J, Ouellet-Oviedo E and Thibault S 2024 Semi-analytical models to engineer a metalens composed of various meta-atoms *J. Opt. Soc. Am.* B **41** 644–52

IOP Publishing

Introduction to Metalens Optics

Ivan Moreno

Chapter 4

Optics of meta-atoms

Meta-atoms or nanoelements are the building blocks of a metalens, which consists of many of them for imparting a phase shift profile to an incident wavefront. Meta-atoms have subwavelength size and spacing to produce a high phase shift despite their thinness. When light passes through a meta-atom, the transmitted light acquires a phase delay, which has a dependence on the shape, size, and orientation of the meta-atom. There are three methods for realizing a full 2π phase shift range. The physical mechanisms mainly involve a geometric (Pancharatnam–Berry) phase, Mie resonance, and propagation phase. The detailed mechanisms of these methods are discussed in this chapter.

Metalenses are constructed by arranging meta-atoms periodically in a 2D plane. In most cases, the unit cells of these ultrathin structures are shorter than wavelengths of interest. Metasurface nanoelements are of subwavelength thickness, have a uniform height (essential for ensuring flatness), and can be batch-fabricated using standard nanofabrication processes, potentially at a low cost. Meta-atoms are mainly devoted to locally controlling the key optical properties of light waves, such as phase, amplitude, and polarization [1]. Understanding how meta-atoms interact with light is fundamental for choosing the appropriate meta-atoms to design a metalens. In this chapter, we explain meta-atoms and their available mechanisms to control phase shifts.

4.1 Principles of phase shift

The principles governing phase shift in optics comprise four main mechanisms: altering the refractive index along the light path, exploiting geometric phase phenomena, utilizing optical resonances, and manipulating total internal reflection (TIR) at different incidence angles. These mechanisms provide diverse means of modulating the phase of light within the range of 0 to 2π. While the first three mechanisms have been extensively studied and applied in metasurfaces and metalenses, the fourth mechanism (TIR) has not yet been implemented in these optical nanostructures. Understanding these mechanisms lays the foundation for advancements in designing and engineering novel metasurface devices with tailored phase profiles, crucial for developments and applications of metalenses.

These four optical mechanisms for changing the phase of light to different values within a range of 0 to 2π can be summarized as follows:

(i) Changing the refractive index along the light path:

This method, also known as dynamic phase or propagation phase, involves altering the refractive index of a material through which light propagates. If a light beam propagates along a path in the space through a medium, it acquires not only a phase from the accumulated path length but also from changes in the refractive index [2]. This method involves altering the refractive index of a material through which light propagates. By varying the refractive index along the light path, the phase of the light wave changes. This can be achieved using materials with electro-optic properties, liquid crystals, gradient refractive index, and the Kerr effect in nonlinear materials, among others. Controlling the refractive index allows for precise manipulation of the phase of the light wave under propagation without changing the propagation distance.

(ii) Geometric phase phenomenon:

The geometric phase, also known as the Pancharatnam–Berry phase, arises from the geometric properties of the light's polarization path on the Poincaré sphere (PS). If the polarization of a light beam is taken along a closed path in the PS (space of polarization states of light), it acquires a phase, which is equal to half the solid angle subtended by the closed path on the PS [3]. Manipulating the polarization state of incident light and altering the enclosed area of the closed path on the PS allows for control over the geometric phase.

(iii) Optical resonance:

In this mechanism, light experiences a phase shift when scattered within optical resonators or nanoantennas. The phase response of a subwavelength antenna (with size smaller than λ) to incident monochromatic and polarized light depends on the antenna size. The scattered light is phase-shifted relative to that of the incident light (excitation). The phase changes with the size of the nanoantenna because in this scale the electric currents in the antenna are not in phase with the incident electromagnetic field [4]. This principle for phase shift can also be understood in the context of strongly confined electric and magnetic multipole modes (figure 4.1). Nanoelements based on resonances can be constructed not only with metallic/plasmonic materials but also with dielectric materials.

(iv) Total internal reflection at different incidence angles:

When light encounters a boundary between two media with different refractive indices, it can undergo total internal reflection (TIR) if the angle of incidence exceeds a critical value. By further increasing the angle of incidence, the phase of the reflected light can be shifted. TIR-based phase modulation is utilized in devices like achromatic prism polarizers [2].

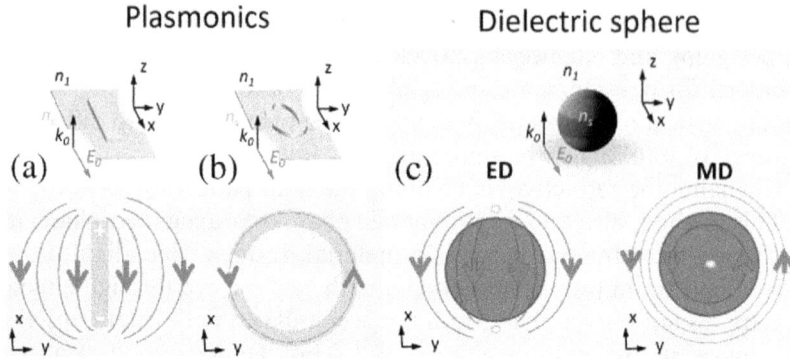

Figure 4.1. Metallic nanoantennas induce plasmonic resonances, and dielectric nanoresonators strongly confine the light within electric and magnetic multipole modes. (a) and (b) show plasmonic resonators. (c) shows a dielectric sphere resonator, with its electric dipole (ED) and magnetic dipole (MD) resonances. (a) Plasmonic rod antennas support only electric resonances. (b) Strong magnetic dipole resonance can be achieved in plasmonic split ring resonators. Image reproduced from reference [5], copyright (2017) Optical Society of America.

Phase shift principles experience a dramatic increment in the nanoscale when light interacts with subwavelength elements like meta-atoms. The current trend, and the most widely implemented phase shift methods are through dielectric meta-atoms. The detailed mechanisms of these methods are discussed in the following sections.

4.2 Metallic meta-atoms

Early advancements in metalenses utilized plasmonic metasurfaces, employing metallic meta-atoms with varied shapes and orientations, commonly referred to as plasmonic antennas. Plasmonics involves the coherent oscillation of electrons at optical frequencies along metal-dielectric interfaces at the nanoscale. When incident light interacts with metallic nanoantennas, free surface electrons synchronize with the electric field, inducing surface electric dipoles. These dipoles generate resonances dependent on material properties, geometry, and natural frequency. By manipulating resonance at these interfaces, subwavelength-sized optical resonators can modulate the phase, amplitude, polarization, and dispersion of visible light [5, 6].

A metallic meta-atom exhibits resonant scattering of light, akin to a radiating metallic antenna, introducing phase shifts via scattering time delays. To explain in simple terms how these tiny antennas introduce a phase shift, let us roughly describe plasmonic antennas by using a single harmonic oscillator model [5, 7]. This simple model represents electronic oscillations by a charge q located at $x(t)$ with mass m on a spring with spring constant k driven by an harmonic incident electric field $E_o e^{i\omega t}$ with frequency ω. Additionally, ohmic losses are considered, introducing a damping coefficient γ. The model to describe the phase response of plasmonic resonance is

$$\frac{d^2x}{dt^2} + \frac{\gamma}{m}\frac{dx}{dt} + \frac{k}{m}x = \frac{q}{m}E_o e^{i\omega t} + \frac{2q^2}{3mc^3}\frac{d^3x}{dt^3}, \qquad (4.1)$$

where there are two damping forces, one is proportional to dx/dt, and the other is due to the radiation reaction that is proportional to d^3x/dt^3. Such a radiation reaction force is the recoil that the charge feels when it emits radiation. Therefore, harmonic motion solution of equation (4.1) yields:

$$x(t) = x_0(\omega)e^{i\omega t} = \frac{AE_o}{\left(\omega_0^2 - \omega^2\right) + i(\omega\Gamma_a + \omega^3\Gamma_s)}e^{i\omega t}, \tag{4.2}$$

where $A = q/m$, $\omega_0 = k/m$, $\Gamma_a = \gamma/m$, and $\Gamma_s = 2q^2/3mc^3$. Note that the phase shift Φ results from the imaginary component in amplitude $x_0(\omega)$, where ω, Γ_a and Γ_s determine the phase. Parameters Γ_a and Γ_s are associated with the non-radiative and radiative damping mechanisms, respectively. The amplitude of the oscillation is in phase with the incident wave, $\Phi = 0$, for $\omega \to 0$. In the other hand, the phase shift is maximum, $\Phi = \pi$, for $\omega \to \infty$. Intermediate phase shift values, $0 < \Phi < \pi$, are obtained for other values of ω, Γ_a and Γ_s. For example $\Phi = \pi/2$ for $\omega = \omega_0$.

However, this scattering phase shift cannot exceed π (see figure 4.2(a)), indicating that plasmonic meta-atoms cannot achieve full phase coverage of 2π using resonant scattering alone. To overcome this limitation, plasmonic resonators are often combined with the Berry–Pancharatnam phase, also known as the geometric phase, which introduces additional phase shifts based on polarized light.

V-shaped nanoantennas (figure 4.2(b)), composed of two joined metallic nano-rods, demonstrated the ability to achieve full phase coverage of 2π for metalenses [4]. The V-antennas combine the plasmonic resonant phase with geometric phase. Within these meta-atoms, different resonant modes are excited by altering the angle of rotation of the antenna, the angle between the two metallic rods, and the direction of the incident light's polarization state (see figures 4.2(c) and (d)). This combination of resonant modes and polarization states allows for a phase coverage of 2π, which is essential for implementing the phase profile of a metalens. Consequently, numerous innovations in plasmonic meta-atoms, metasurfaces and metalenses have been explored [8–10]. Plasmonic metalenses exhibit polarization dependence, typically operating with circularly polarized light. Their applications are specialized, featuring unique functionalities such as switching a focusing lens to a diverging lens when the incident light's polarization is converted into the opposite polarization. Because of their low aspect ratio or small subwavelength thickness, fabrication of plasmonic meta-atoms can be simpler than that of dielectric meta-atoms. However, plasmonic metalenses have limited theoretical focusing efficiency due to inherent Ohmic losses in metals (conductors). Therefore, current efforts are focused on developing high-efficiency, low-loss dielectric metalenses [11, 12].

4.3 Dielectric meta-atoms

In recent years, the efficiency of optical metalenses has increased significantly by switching from metallic meta-atoms to their dielectric counterparts [11, 12]. Early metalenses used ultrathin metallic nanoelements but because of the intrinsic absorption losses and fundamental efficiency limits of metals, recent metalenses usually use thicker dielectric meta-elements. Dielectric meta-atoms are characterized by their refractive

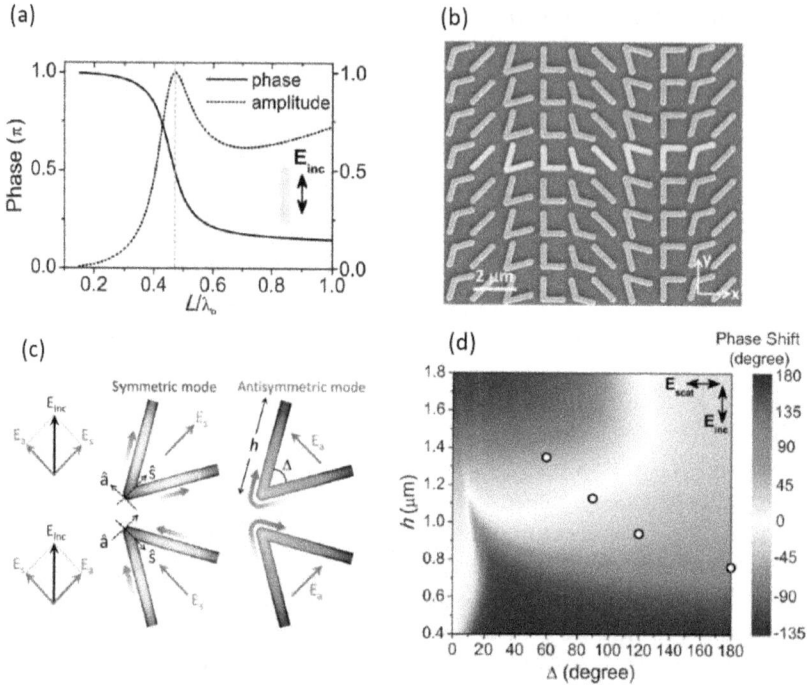

Figure 4.2. Plasmonic meta-atoms. (a) Theoretical phase shift and amplitude of scattered light from a simple metallic rod nanoantenna, where L is the rod length. (b) SEM image of the fabricated gold antenna on a silicon wafer. (c) Schematic of the the V-shaped antenna, showing leg size h, V-angle Δ, and incident polarization of light E_{inc}. (d) Analytically calculated phase shift of the scattered light by gold V-antennas for various lengths h and V-angles Δ at $\lambda = 8$ mm. Image reproduced from reference [4] with permission from AAAS.

index, and then they are thicker to achieve a phase coverage of 0–2π. Dielectric metalenses have potential for high efficiency, low cost, and a planar and thin form factor.

Despite different types of dielectric meta-atoms giving rise to essentially the same macroscopic thickness of the metalens, there are numerous optical differences among them. Table 4.1 contrasts the peculiarities of the main types, showcasing that the optimal performance of each meta-atom type depends on the specific application requirements. While all types of dielectric nanoelements are desired to be made of materials with a high refractive index in the spectral range of interest, many metalenses have been developed with a low contrast refractive index. Additionally, all kinds of dielectric meta-atoms can exhibit undesired Fabry–Pérot resonances, although these resonances can be exploited for certain applications [13, 14].

While meta-atoms with simple geometric shapes such as nanorods, nanofins, and nanodisks have seen widespread use due to their ease of fabrication, many studies have explored metalenses composed of different types of dielectric nanoelements. Nevertheless, practical metalenses typically employ cylindrical (nanopost) meta-atoms with larger aspect ratios. In this section, our focus lies on dielectric metasurfaces, with an emphasis on explaining the phase modulation mechanisms of dielectric meta-atoms.

Table 4.1. Some optical properties of three main dielectric meta-atoms or nanoelements.

Property	'Propagation' phase	'Resonant' phase	'Geometric' phase
Geometry	Rods	Pills	Fins
Main modulation parameter	Diameter	Diameter	Orientation (angle)
Polarization	Not sensitive	Low sensitive	Sensitive
Multiwavelength operation	Low sensitive	Sensitive (monochromatic)	Not sensitive (highly achromatic)
Effective transmission	High	Very high	Medium

4.3.1 Propatation phase

A propagation phase meta-atom is a dielectric nanopillar that functions as a truncated or short waveguide, introducing phase shifts through light propagation, with negligible absorption loss [15, 16]. Functioning as truncated waveguides, they are scatterers with low-quality factor resonances, with phase accumulation achieved via propagation. However, Fabry–Perot-like resonances are inevitable due to reflections at the nanorod ends, caused by the change in refractive index at these points. Additionally, a small portion of the incident light is scattered, leading to the imparting of phase into transmitted light through the superposition of propagation, resonance, and radiation effects. Nevertheless, the dominant phase mechanism arises from the accumulated phase during propagation.

Nanopillars of varying diameters create different effective refractive indices, resulting in different phase shifts (figure 4.3(a)). Taller nanopillars (height $H \sim \lambda$) enable a full 0–2π range of phase shifts with a smaller variety of pillar sizes, offering more flexibility to enhance their optical performance, for example avoiding coupling effects with neighbor meta-atoms. Propagation phase meta-atoms come in various shapes, including rectangular, cylindrical, truncated cone, and others (figure 4.3(b)).

The propagation phase is controlled by managing the phase retardation through the meta-atom, which acts as a nanowaveguide. The phase retardation induced by the nanopillar can be approximated as

$$\Phi \approx \frac{2\pi}{\lambda} n_{\text{eff}} H, \tag{4.3}$$

where λ is the wavelength of incident light, n_{eff} the effective refractive index ($n_{\text{eff}} < n$, where n is the nanopillar refractive index), and H the meta-atom height (propagation length). Typically, the effective refractive index is a function of the nanopillar volume, but since most metalenses are flat (with constant H), n_{eff} becomes a function of the cross-section area or diameter. Larger-diameter meta-atoms result in a larger n_{eff}, leading to a greater phase shift.

Nanoposts can modulate the phase of propagating light based on the meta-atom size. An essential property of nanopillars is that their phase shift remains independent

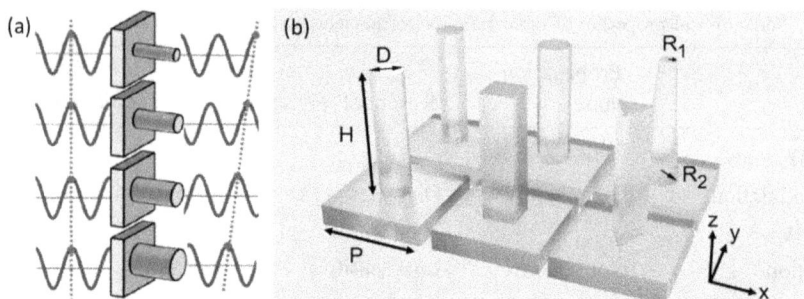

Figure 4.3. Propagation phase meta-atoms. (a) Phase shifts for different diameters. Reproduced from [17] CC BY 4.0. (b) Schematic of several nanorods with several shapes. Figure illustrates cross-section shapes with rotational symmetry larger than C3: triangle (C3), square (C4), hexagon (C6), octagon (C8), and C-infinite (circular and truncated cone). Meta-atoms of height H, and unit cells in a square lattice of dimensions $P \times P$. Reprinted with permission from [1]. Copyright (2024) American Chemical Society.

of their geometric shape for geometric symmetries larger than C3 [1]. This invariant optical property simplifies the metalens design and manufacturing process when using nanoposts. Nanorods with geometric symmetries larger than C3 include cylinders, triangles, squares, hexagons, octagons, and truncated cones (figure 4.3(b)).

Generally, nanopillars are cylindrical posts, making them polarization-independent due to its symmetrical shape. However, all other geometries shown in figure 4.3(b) are also polarization-independent [1]. Thus, metalenses employing nanoposts can produce images under unpolarized light. Ideally, the transmission through a nanopillar remains constant across the phase shift modulation. However, Fabry–Perot resonances are unavoidable, affecting its transmittance, which also depends on nanopost diameter. Therefore, meta-atoms without resonance within the selected diameter range are desired for most metalens designs. As the phase shift is introduced through propagation distance, these meta-atoms act as truncated waveguide elements operating in a broadband regime with high transmittance and full phase modulation. However, the phase shift is somewhat sensitive to wavelength operation, requiring careful meta-atom design for chromatic correction in metalens applications.

To maximize the efficiency of the meta-atom, the height (H) of the nanopillars and the unit cell size (P) need to be optimized at the design wavelength (figure 4.3(b)). Since the phase shift relies on the waveguiding effect, H must be sufficiently tall to cover a 2π phase range across the nanopillar diameter (D) range. While the minimum diameter is constrained mainly by fabrication limitations, the maximum D is dictated by the unit cell size [15]. Therefore, the meta-atom design process begins with the designated design wavelength (λ), followed by the exploration of optimal H and P values. Typically, this involves simulations to generate phase and transmission maps as functions of D and H, considering various P values. The goal is to identify the nanopost height value that enables achieving full phase coverage (2π) with high transmission by adjusting nanopost diameters (figure 4.4(a)). Subsequently, the optimal range of nanorod diameters is employed to design and fabricate the metalens (figure 4.4(b)).

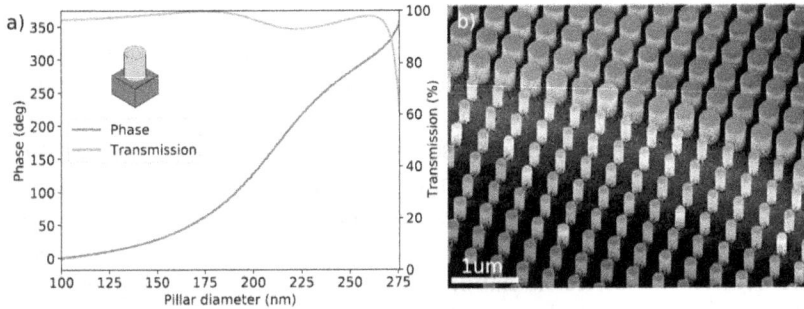

Figure 4.4. Cylindrical nanopillars. (a) Simulated transmitted phase (blue line, left-hand axis) and transmitted power (orange line, right-hand axis) for amorphous silicon nanoposts with different diameters, for $H = 500$ nm and $P = 400$ nm. (b) Scanning-electron microscope (SEM) image of a manufactured metalens with such nanopillars in a square lattice. Images reproduced from reference [18] CC BY 4.0.

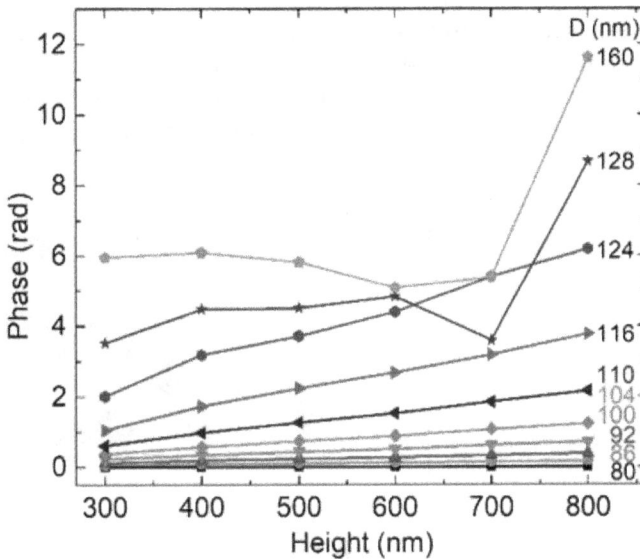

Figure 4.5. Phase shift vs. meta-atom height H, for cylindrical silicon meta-atoms. Graphs for several different cylinder diameters D, from 80 to 160 nm. This shows the validity of equation (4.3), for this case it is valid for $D < 124$ nm. Reprinted with permission from [1]. Copyright (2024) American Chemical Society.

Additionally, equation (4.3) is typically considered an equality, but it is approximate. The phase Φ should show a linear variation as the height H increases, but this is valid only for a range of nanopost diameters [1]. This behavior is illustrated in figure 4.5, depicting the phase shift of silicon cylindrical nanoposts vs. height H for several diameters D of the cylinders.

4.3.2 Resonant phase

This type of dielectric nanoelement is ultrathin (thickness $H \ll \lambda$). Also known as dielectric nanoresonator or Huygens' meta-atom, it introduces phase shifts related to strong excited Mie-type scattering resonances [19, 20]. Unlike its metallic counterpart, a dielectric resonant phase meta-atom achieves a 2π phase coverage by using both the magnetic dipole (MD) and electric dipole (ED) resonances (figure 4.1(c)). The phase shift can be tuned by adjusting the nanoelement dimensions and shapes [20], allowing a maximum of 2π phase shift with near-unity transmission for the designed wavelength. Their simplest shape is the nanodisk (figure 4.6(a)).

Resonant phase metasurfaces have been conventionally termed as Huygens' metasurfaces, owing to the analogy between the Huygens' point sources and the scattering behavior of dielectric nanoresonators. According to Huygens's Principle [2], waves propagate as if the wavefronts were composed of an array of point sources, each emitting a spherical wave in forward direction but not backward. Resonant phase meta-atoms are designed to scatter light forward but not backward. This backward radiation cancellation is achieved through the so-called first Kerker condition. Such a condition is achieved when ED and MD dipoles are oscillating in phase, creating a minimum in the backscattered light. Ideally, a subwavelength array of nanodisks with ED and MD resonances of equal strength and width show unitary transmission under plane-wave illumination. However, resonant dielectric nanoelements operate efficiently only within a narrow bandwidth. Additionally, the resonance mode coupling between adjacent meta-atoms may lead to phase profile

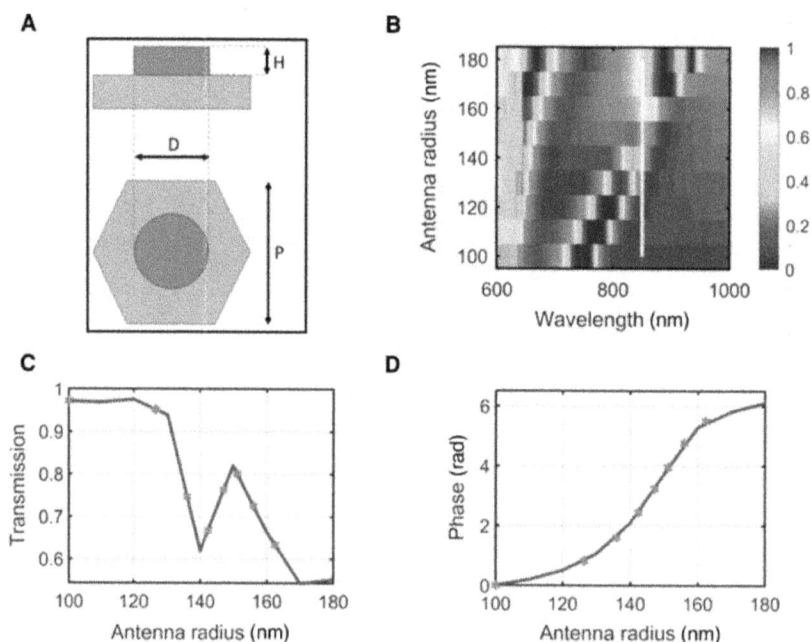

Figure 4.6. Resonant phase meta-atoms. (A) Nanodisk unit cell, where P is the unit cell size, H is the disk height, and D is the disk diameter. (B) Transmission of a Huygens a-Si nanodisk as a function of wavelength and disk radius, where the vertical white line shows the locus of values for graphs (C) and (D). (C) Transmission as a function of nanodisk radius. (D) Phase shift as a function of nanodisk radius. Image reproduced from reference [21] CC BY 4.0.

errors, degrading the metalens focusing performance in some cases. Therefore, Huygens metalenses are usually designed for monochromatic operation, but can support high numerical aperture and wide field of view (FOV). For example, a moderately wide-FOV Huygens metalens was implemented with nanodisks for outdoor imaging applications in near-IR over a relatively broad spectral range [21]. Therefore, resonant phase metalenses and metasurfaces are constantly under extensive research for improvement.

To achieve the required phase shifts for the metalens phase profile with resonant meta-atoms, finite-difference time-domain (FDTD) simulations must be performed. The nanoresonators are made of a chosen material, usually on a glass substrate. The unit cell size or lattice period must be chosen considering subwavelength size and phase sampling requirements (see more in chapter 5). The lattice period should not be too small to minimize nanoelements coupling, which may affect the Kerker condition. To achieve the Kerker condition, a simulation scanning over the nano-antenna geometric parameters and wavelengths must be performed, while computing the transmission. Then the design wavelength is selected for maximum transmission through the selected range of geometric parameters (e.g., nanodisk diameters). Figure 4.6(b) shows an example of such simulations for nanodisks at a design wavelength of 850nm when the simulated antenna height is 140 nm [21].

In the case of nanodisks, to find the optimal meta-atom dimensions, an additional simulation scanning over the nanodisk radius and height must be performed, while computing the transmission and phase shift. In the case of free-form Huygens meta-atoms, a simulation scanning over the cross-section shapes and sizes must be performed, while computing the transmission and phase shift. An example of transmission and phase responses of nanodisks is shown in figures 4.6(c) and (d), for nanodisk radii from 100 to 180 nm [21]. These simulations used a hexagonal lattice with a period of 500 nm, a design wavelength of 850 nm, and an optimal nanodisk height of 140 nm.

Additionally, the optical properties of resonant meta-atoms can be modulated by changing their geometry and material properties. For example, Huygens nanoelements can shift the phase between two values in a reconfigurable varifocal metalens [20], allowing incident light to focus on one focal point when meta-atoms are in the amorphous state and on a different focal point when in the crystalline state. Figure 4.7(a) illustrates such meta-atoms operating at a wavelength of 5.2 μm, constituting an experimentally demonstrated non-mechanical tunable metalens. A selection of 16 optimal meta-atom designs with various regular geometries, such as 'H' and '+' shapes (figure 4.7(b)), was obtained by sweeping the geometric parameters through simulations for two optical states (amorphous and crystalline). The optimal meta-atom designs exhibited maximum transmission and the ability to replicate the metalens phase profile.

4.3.3 Geometric phase

Geometric phase, also referred to as Pancharatnam–Berry (PB) phase, in nanoelements, or meta-atoms, are also known as anisotropic meta-atoms or nanofins,

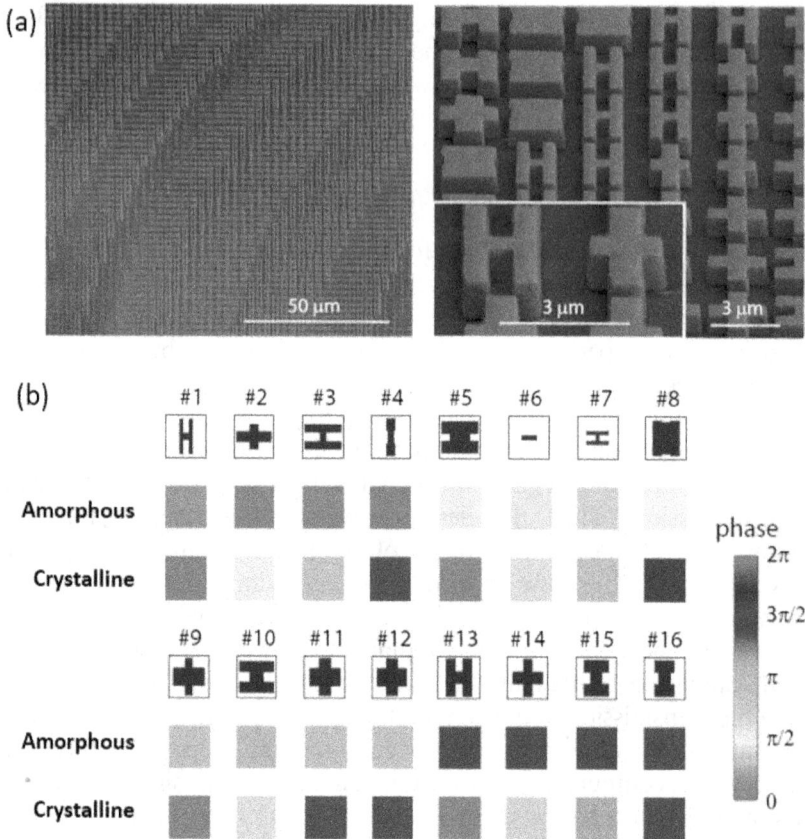

Figure 4.7. Huygens meta-atoms with various geometries. (a) SEM images of the meta-atoms on a reconfigurable metalens at two scales. (b) 16 optimal nanoelement designs of a pool with various 'H' and '+' geometries. Color squares indicate the phase values for two different optical switchable states: amorphous and crystalline. Image reproduced from reference [20] CC BY 4.0.

typically possessing a rectangular shape. These building blocks execute phase control solely by adjusting their orientation angle. PB-phase meta-atoms achieve complete phase coverage by altering the orientation angle of identical meta-atoms across the metalens (figure 4.8). The rotation angle θ of these meta-atoms affects the phase of incident circularly polarized light under transmission. If incident waves are right-handed circularly polarized, the rotation angle θ yields a phase shift of 2θ, accompanied by a polarization conversion to left-handed circular polarization. This phenomenon, known as the PB phase or geometric phase, is theoretically demonstrated to be equivalent to twice the rotation angle of anisotropic nanoelements.

Therefore, a nanofin-based metalens has the advantage of a simple meta-atom design and fabrication. The metalens phase profile is implemented via rotation of each meta-atom at a given coordinate (x, y) by an angle $\theta(x, y)$. For example, in the case of a hyperbolic phase profile, employing right-handed circularly polarized incident light necessitates rotating each nanofin at radial position (r) by an angle [22]:

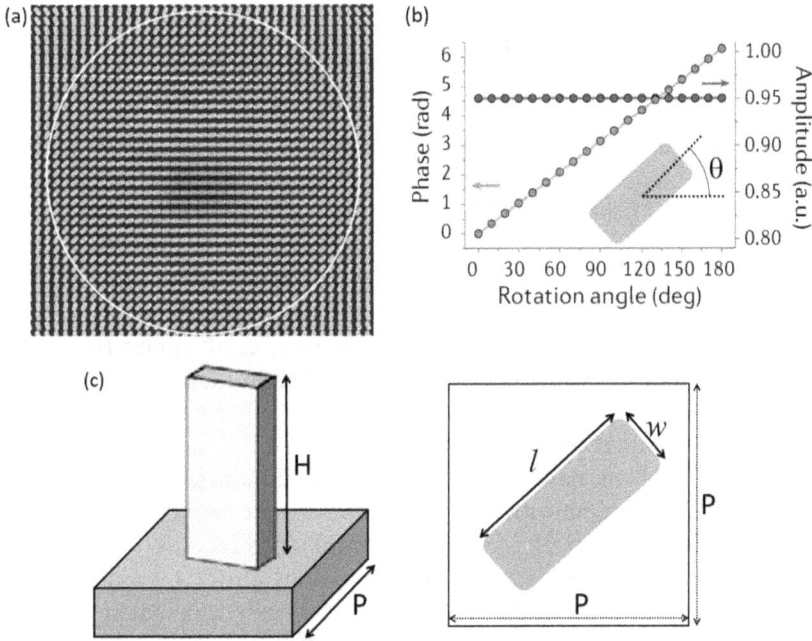

Figure 4.8. Geometric phase meta-atoms. (a) SEM image of a metalens with nanofins. From [22], reprinted with permission from AAAS. (b) Transmitted phase shift and amplitude due to a nanofin, in function of its rotation angle θ for incident circularly polarized light. (c) Nanofin size: long side (l), short side (w), height (H), unit cell size (P). (b) and (c) Adapted by permission from Springer Nature Service Centre GmbH [Nature] [Nature Reviews Materials] [23], copyright (2020).

$$\theta(r) = -\frac{k}{2}(\sqrt{r^2 + f^2} - f), \tag{4.4}$$

A metalens with nanofins enables intuitive design of nanoelements since the phase shift of transmitted light is solely a function of the rotation angle θ. However, nanofins have the limitation of operating solely under circularly polarized light. The operation of PB-phase meta-atoms depends on different phase shifts depending on the polarization state along the long (l) and short (w) axis of the nanofin (figure 4.8(c)). This is equivalent to have a birefringence arising from the asymmetric cross-section of nanofins, which requires an appropriate design of width, length, and height.

To maximize their efficiency, the nanofins should operate as half-waveplates. In other words, to maximize polarization conversion efficiency, a nanofin should behave as an ideal half-waveplate, which is achieved by selecting its length (l) and width (w) such that the phase difference between polarizations along the l and w axes is π [23].

The optical behaviour of a nanofin can be explained as that of a rotated waveplate. For circularly polarized incidence $\boldsymbol{E}_{\text{in}} = [1 \ \pm i]^T$ (where T indicates matrix transpose, and the ($-$) sign denotes left-handed and ($+$) right-handed polarized light), the transmitted electric field $\boldsymbol{E}_{\text{out}}$ from a θ rotated nanofin is [23–25]:

$$E_{\text{out}} = \frac{t_l + t_w}{2}\begin{pmatrix} 1 \\ \pm i \end{pmatrix} + \frac{t_l - t_w}{2}\exp\left(\pm i2\theta\right)\begin{pmatrix} 1 \\ \mp i \end{pmatrix}, \tag{4.5}$$

where t_l and t_w represent complex transmission coefficients for the electric field of incident light polarized along the long l and short w axes of the nanofin, respectively. These coefficients involve the propagation phase. The angle θ is defined as the counterclockwise rotation angle of the meta-atom with respect the x-axis. According to equation (4.5), rotating the nanofin by θ results in a phase shift of 2θ, illustrating the origin of PB phase. Therefore, full phase modulation can be tuned from 0 to 2π by rotating the nanoelement from 0 to π. Equation (4.5) considers the meta-atom under normal monochromatic plane-wave incidence, establishing the relationship between incident and transmitted fields using a rotation matrix $R(\theta)$ and the Jones matrix [6, 17].

The first term of equation (4.5) represents unwanted light (with no added phase shift among all metalens nanofins), which can be minimized if the meta-atom is designed as a half-waveplate with a π phase difference between polarizations along l and w axes. In this scenario, the first term is minimized by $\text{abs}(t_l+t_w) = \text{abs}(t_l) - \text{abs}(t_w)$, while the second term is maximized by making $\text{abs}(t_l-t_w) = \text{abs}(t_l) + \text{abs}(t_w)$. This leads to maximal polarization conversion efficiency, calculated as the ratio of transmitted optical power with opposite polarization helicity to the total incident power. The first term of equation (4.5) is the co-polarized component, and the second term is the crossed-polarized component with the geometric phase shift. The second term acquires opposite polarization handedness, while the first term maintains its polarization. The sign of the phase shift depends on the incident circular polarization (incident light with right-handed circular polarization undergoes a negative phase). Because the sign of the geometric phase shift depends on the polarization helicity of the incident light, metalenses with nanofins are intrinsically polarization-sensitive. Several special designs aim to minimize polarization sensitivity, for example by limiting nanofin rotations to $\theta = 0°$ and $90°$ [25].

The main advantage of PB-phase meta-atoms is that the phase shift 2θ is wavelength-independent, enabling the metalens to ideally operate across a broad bandwidth, facilitating broadband imaging. Consequently, rotating identical PB-phase meta-atoms achieves broadband 2π phase coverage. However, it must be noted that t_l and t_w are wavelength-dependent complex numbers, and thus, the length and width of the nanofin are typically tailored for a specific design wavelength λ.

In addition, metalenses with nanofins can only focus light when the incident light is circularly polarized. In the ideal case, 100% polarization conversion efficiency is possible by achieving the complete conversion of circularly polarized light. However, due to coupling effects between unpolarized light and circularly polarized light, in general imaging applications, the ideal efficiency is 50%, corresponding to the ideal transmission efficiency of a circular polarization polarizer for unpolarized incident light, which is 50%.

Ongoing research aims to enhance broadband operation and polarization insensitivity. Approaches include combining other phase mechanisms like propagation phase and resonant phase [25, 26]. In cases involving propagation and geometric phase, nanofin rotation angles and sizes are adjusted simultaneously. Other methods involve employing complex meta-atom shapes and meta-molecules.

4.4 Meta-atom materials

Meta-atoms use light–matter interactions that are determined by their material properties and the properties of interacting light. Thus, understanding these properties is crucial for the intuitive design of metalenses. Broadly speaking, two types of materials are employed: metallic and dielectric. Metallic meta-atoms exploit plasmonic resonances, while dielectric ones confine and scatter light through their subwavelength structures. While most high-efficiency metalenses utilize dielectric meta-atoms, research on reflective or specialized metallic metalenses continues to progress [6].

Meta-atoms can be designed for various target wavelengths depending on the application, with material selection dictated by the spectral range. The optical properties of a material are characterized by its complex refractive index ($n+ik$), where the real part (n) correlates with the speed of light, and the imaginary part (k) corresponds to light absorption within the material. A higher n value results in a larger phase shift, while a higher k value leads to increased absorption loss. Therefore, materials with high n and low k values are typically preferred. Materials used for fabricating metalenses span from metals and metal-insulator-metal structures to dielectrics.

Early optical metalenses utilized plasmonic materials, but their high optical losses and low forward-scattering properties led to a shift toward dielectric materials. Currently, high-index dielectric materials are the preferred choice to construct meta-atoms. Because dielectrics are naturally compatible with current metal-oxide semi-conductor (CMOS) technologies, dielectric-based metalenses have advantages in most applications compared to their metallic counterparts [27].

Among the popular materials for metalenses are titanium dioxide (TiO_2), gallium nitride (GaN), and amorphous silicon (a-Si). Amorphous silicon is a common choice for metalenses due to its high refractive index, low loss, and manufacturing compatibility with CMOS technology. Amorphous TiO_2 and GaN have refractive indices in the range of 2.0–2.4, providing high contrast and light confinement. TiO_2 meta-atoms exhibit low surface roughness, are completely transparent in the visible range, allow the fabrication of nanoelements with vertical walls, and a high refractive index ($\sim$2.4) [22]. Several dielectric meta-atoms meet the key requirements of metalenses: low absorption, high refractive index, and ease of manufacture [28]. Currently, TiO_2 [22], GaN [29], crystalline silicon (c-Si) [30], and silicon nitride (Si_3N_4) [31] are often used in the visible regime due to their high transmission at visible wavelengths. In the other hand, c-Si has a high refractive index ($n \sim 4$) and a relatively low absorption in the visible range (absorption coefficient $\sim$ 0.05). In summary, meta-atoms should be subwavelength in scale and composed of low-optical loss dielectrics with a high index of refraction, easily manufacturable with the required manufacturing resolution in the nanoscale. Additionally, the typical substrate is fused silica, with meta-atoms commonly positioned on a glass or SiO_2

substrate, although other substrates like gold substrates can be employed depending on the application [27].

4.5 Problems

4.1 **Design of Plasmonic Meta-Atoms.** Deduce the phase shift introduced by the simple model given by equation (4.2). Discuss how the optical properties of the metal influence the phase shift and how this can be optimized.

4.2 **Dielectric Meta-Atoms and Propagation Phase**. Analyze a dielectric meta-atom that achieves a desired phase shift through propagation phase. Given a specific dielectric material with refractive index n, derive the minimum thickness of the meta-atom required to achieve a phase shift of 2π at a wavelength λ. Discuss the conditions to get the minimum and maximum phase shifts by varying the effective refractive index or nanopillar diameter.

4.3 **Resonant Phase in Dielectric Meta-Atoms.** Consider a dielectric meta-atom that uses resonant phase to achieve a specific phase shift at a wavelength λ. Discuss the role of the resonant frequency and quality factor Q to achieve a desired phase shift.

4.4 **Optimization of Phase Shift Efficiency.** Consider a metalens designed using dielectric meta-atoms to achieve a specific phase shift for incident light at a wavelength λ. The metasurface is designed to provide a phase shift through the three mechanisms: propagation phase, resonant phase, or geometric phase. Develop a strategy to optimize the phase shift efficiency for this metalens composed of a combination of these three types of phase-shifting meta-atoms. Define a metric for the efficiency of the phase shift as a function of the parameters involved, and propose a method to maximize the overall efficiency of the metalens.

Bibliography

[1] Basilio-Ortiz J C and Moreno I 2024 Unveiling invariant optical properties of dielectric meta-atoms *Nano Lett.* **24** 4987–92
[2] Hecht E 2002 *Optics* (San Francisco, CA: Addison-Wesly)
[3] Gutiérrez-Vega J C 2011 Pancharatnam–Berry phase of optical systems *Opt. Lett.* **36** 1143–5
[4] Yu N, Genevet P, Kats M A, Aieta F, Tetienne J-P, Capasso F and Gaburro Z 2011 *Science* **334** 333
[5] Genevet P, Capasso F, Aieta F, Khorasaninejad M and Devlin R 2017 *Optica* **4** 139
[6] Moon S-W *et al* 2022 Tutorial on metalenses for advanced flat optics: design, fabrication, and critical considerations *J. Appl. Phys.* **131** 091101
[7] Kats M A, Yu N, Genevet P, Gaburro Z and Capasso F 2011 Effect of radiation damping on the spectral response of plasmonic components *Opt. Express* **19** 21748–53
[8] Meinzer N, Barnes W L and Hooper I R 2014 Plasmonic meta-atoms and metasurfaces *Nat. Photonics* **8** 889–98
[9] Chen X, Huang L, Muhlenbernd H, Li G, Bai B, Tan Q, Jin G, Qiu C W, Zhang S and Zentgraf T 2012 Dual-polarity plasmonic metalens for visible light *Nat. Commun.* **3** 1198
[10] Zhang J, ElKabbash M, Wei R, Singh S C, Lam B and Guo C 2019 *Light Sci. Appl.* **8** 53

[11] Kamali S M, Arbabi E, Arbabi A and Faraon A 2018 A review of dielectric optical metasurfaces for wavefront control *Nanophotonics* **7** 1041–68

[12] Arbabi A and Faraon A 2023 Advances in optical metalenses *Nat. Photon.* **17** 16–25

[13] Basilio-Ortiz J C and Moreno I 2022 Multilayer dielectric metalens *Opt. Lett.* **47** 5333–6

[14] Basilio-Ortiz J C and Moreno I 2023 All dielectric reflective metalens based on multilayer meta-atoms *Opt. Lett.* **48** 5647–50

[15] Khorasaninejad M, Zhu A Y, Roques-Carmes C, Chen W T, Oh J, Mishra I, Devlin R C and Capasso F 2016 Polarization insensitive metalenses at visible wavelengths *Nano Lett.* **16** 7229–34

[16] Khorasaninejad M and Capasso F 2017 Metalenses: versatile multifunctional photonic components *Science* **358** eaam8100

[17] Kim J, Yang Y, Badloe T, Kim I, Yoon G and Rho J 2021 Geometric and physical configurations of meta-atoms for advanced metasurface holography *InfoMat* **3** 739–54

[18] Egede Johansen V *et al* 2024 Nanoscale precision brings experimental metalens efficiencies on par with theoretical promises *Commun. Phys.* **7** 123

[19] Decker M *et al* 2015 High-efficiency dielectric Huygens' surfaces *Adv. Opt. Mater.* **3** 813–20

[20] Shalaginov M Y *et al* 2021 Reconfigurable all-dielectric metalens with diffraction-limited performance *Nat. Commun.* **12** 1225

[21] Engelberg J, Zhou C, Mazurski N, Bar-David J, Kristensen A and Levy U 2020 Near-IR wide-field-of-view Huygens metalens for outdoor imaging applications *Nanophotonics* **9** 361–70

[22] Khorasaninejad M *et al* 2016 Metalenses at visible wavelengths: diffraction-limited focusing and subwavelength resolution imaging *Science* **352** 1190–4

[23] Chen W T, Zhu A Y and Capasso F 2020 Flat optics with dispersion-engineered metasurfaces *Nat. Rev. Mater.* **5** 604–20

[24] Pan M, Fu Y, Zheng M, Chen H, Zang Y, Duan H *et al* 2022 Dielectric metalens for miniaturized imaging systems: progress and challenges *Light: Sci. Appl.* **11** 195

[25] Chen W T, Zhu A Y, Sisler J, Bharwani Z and Capasso F 2019 A broadband achromatic polarization-insensitive metalens consisting of anisotropic nanostructures *Nat. Commun.* **10** 355

[26] Mueller J B, Rubin N A, Devlin R C, Groever B and Capasso F 2017 Metasurface polarization optics: independent phase control of arbitrary orthogonal states of polarization *Phys. Rev. Lett.* **118** 113901

[27] Sun S, He Q, Hao J, Xiao S and Zhou L 2019 Electromagnetic metasurfaces: physics and applications *Adv. Opt. Photon* **11** 380–479

[28] Liang H, Martins A, Borges B-H V, Zhou J, Martins E R, Li J and Krauss T F 2019 High performance metalenses: numerical aperture, aberrations, chromaticity, and trade-offs *Optica* **6** 1461–70

[29] Chen B H *et al* 2017 GaN metalens for pixel-level full-color routing at visible light *Nano Lett.* **17** 6345–52

[30] Liang H *et al* 2018 Ultrahigh numerical aperture metalens at visible wavelengths *Nano Lett.* **18** 4460–6

[31] Fan Z-B, Shao Z-K, Xie M-Y, Pang X-N, Ruan W-S, Zhao F-L, Chen Y-J, Yu S-Y and Dong J-W 2018 Silicon nitride metalenses for close-to-one numerical aperture and wide-angle visible imaging *Phys. Rev. Appl.* **10** 014005

Chapter 5

Optical design of metalenses

Welcome to the last chapter of this introduction to metalens optics. This chapter provides an introductory exploration of the advanced process involved in designing metalenses, an emerging technology with transformative potential in optics. This chapter aims to provide a detailed and basic guide for designing metalenses, covering key optical principles, basic steps, considerations, and design methods to achieve optimal performance. We embark on this journey by outlining the basic design process, followed by a discussion on the numerical aperture (NA) and a broader explanation of the significance of metalens phase profiles. Furthermore, we explain the important role of arrays of meta-atoms, exploring different geometries and the associated numerical aperture limits. Chromatic aberrations and focusing efficiency, key challenges in metalens design, are also addressed. Through this chapter, readers will gain valuable insights into the basic principles and advanced techniques essential for designing high-performance metalenses for diverse optical applications.

5.1 Basic design process

Because a metalens consists of thousands or millions of meta-atoms, which span a wide range of sizes, orientations, or shapes, designing a metalens is one of the most challenging tasks in optics. The design requires several steps including calculating the required phase profile, designing the meta-atoms needed to physically implement it, and testing the metalens performance [1–3].

A schematic of the basic design process is shown in figure 5.1. This procedure consists of three steps: (a) to get the desired metalens phase profile, usually calculated through an analytical expression, which depends on the application such as beam focusing, imaging, and steering; (b) to obtain a library of meta-atoms (chosen group of nanoelements) with their corresponding phases in function of a physical parameter (for example, nanopillar diameter) through electromagnetic simulations; (c) to approach the target phase profile through meta-elements by using

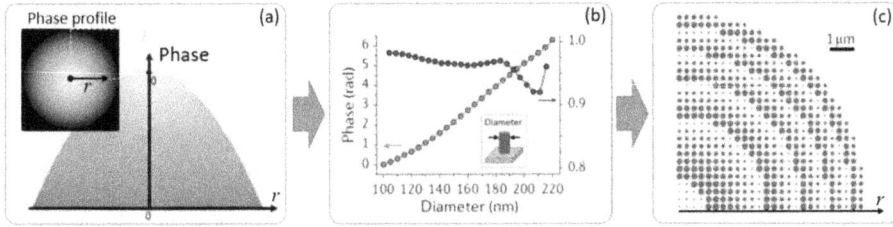

Figure 5.1. A flow chart for designing a basic metalens in three steps. (a) Defining or obtaining the desired phase profile. (b) Calculating a suitable phase shift meta-atom response vs. meta-atom parameter (e.g. diameter) by electromagnetic simulations. (c) To obtain the nanostructure spatial distribution for all nanoelements across the metalens surface. Adapted by permission from Springer Nature Service Centre GmbH [Nature] [Nature Reviews Materials] [1], copyright (2020).

the library to determine the spatial distribution for all meta-atoms across the metalens.

The metalens design includes the building of a library about the phase shift and light transmission in function of the materials and geometries of meta-atoms, which should include a full phase modulation (2π) with a suitable optical transmission. Using both the meta-atom library and target phase profile, metalenses are implemented by selecting from the library the meta-atom that approaches the phase profile at each position on the metalens surface.

The design approach depicted in figure 5.1 is a basic design approach that depends on a pre-determined phase profile. It is a basic approach because it may be sensitive to changes in the incident conditions, i.e., the phase shift introduced by the meta-atoms may depend on frequency, angle of incidence, polarization of light, and so on. Therefore, designing a metalens with high performance for a range of incident conditions is an additional challenging task, which includes the design of achromatic, large field of view, and high numerical aperture metalenses.

5.2 Numerical aperture (NA)

The amount of light gathered by a lens is one of the factors that determines the brightness of the image [4]. The numerical aperture and the f-number are the parameters used for describing this optical property. The f-number is a useful parameter when the object is far away, for example in telescopes. The NA is more appropriate when the object is near, for optics at finite conjugates when object and image distances are both finite, for example in microscopy. The NA indicates the ability of a lens to focus and collect light, and also to resolve small features. The NA is defined as

$$\mathrm{NA} = n_i \sin\theta_{\max}, \tag{5.1}$$

where $\theta_{\max}$ represents the maximum acceptance angle of the lens. The angle $2\theta_{\max}$ corresponds to the vertex angle of the largest cone of rays that can enter the lens, given by $\sin\theta_{\max} = D/2f$, where D and f are the diameter and focal distance of the lens. The square of NA is a measure of the light-gathering power of the optical system. And the relation between the f-number and NA is given by $f/\# = 1/2(\mathrm{NA})$.

Increasing the NA increases the collected light but also increases the gradient of the phase profile.

5.3 Metalens phase profiles

Ultimately, the phase profile that a metalens imparts to the incident light determines the desired optical characteristics of the metalens. The hyperbolic function (equation (2.12)) serves as a basic phase profile for metalenses due to its absence of spherical aberration. However, other aberrations such as comatic, astigmatic, and chromatic aberrations may still be present. Correcting or eliminating any of these aberrations can enhance the resulting imaging or focusing performance [4, 5]. Some aberrations can be reduced by simply adjusting the shape of the metalens phase profile, while others may require more extensive modifications to the entire metalens system [6].

For example, a high-NA metalens with a hyperbolic phase profile exhibits coma and other off-axis aberrations, limiting its field of view. The off-axis aberrations may be reduced by imposing a phase-correcting polynomial function, given by [6, 7]:

$$\Phi(r) = -k(\sqrt{r^2 + f^2} - f) + \sum_{n=1}^{5} a_n \left(\frac{r}{R}\right)^{2n}, \tag{5.2}$$

where k and f are the wave number and focal length, respectively, a_n are optimization coefficients, and R is the metalens radius. The phase profile given by equation (5.2) is a numerically optimized function. An example of such a phase profile is shown in figure 5.2(a), for $\lambda = 532$ nm and $f = 342$ μm. The optimization coefficients a_n were obtained through ray tracing optimization [7], such that all rays for various incidence angles (up to 25°) fall within small focal spots (diffraction-limited Airy disks). Further correction of off-axis aberrations may be achieved by introducing an additional metalens with a polynomial phase function, given by [7]:

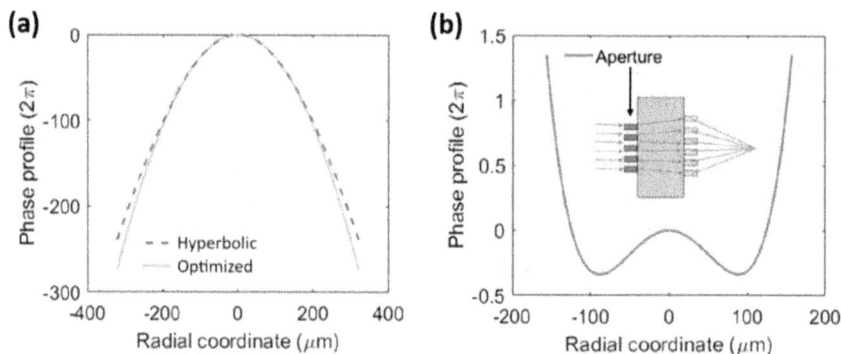

Figure 5.2. Phase profiles different from the basic hyperbolic profile. (a) Comparison of hyperbolic and an optimized phase profile to reduce off-axis aberrations. It shows the phase profile of equation 5.2. (b) Phase profile of an aperture metalens in a metalens doublet (shown in inset) for correcting off-axis aberrations. It shows the phase profile of equation (5.3). Reprinted with permission from [7], copyright (2017) American Chemical Society.

$$\Phi(r) = \sum_{n=1}^{5} b_n \left(\frac{r}{R_a}\right)^{2n}, \qquad (5.3)$$

where b_n are optimization coefficients, and R_a is the radius of the additional metalens (aperture metalens). An example of such a phase profile is shown in figure 5.2(b), which, when combined with the metalens exhibiting the optimized phase profile shown in figure 5.2(a), corrects off-axis aberrations in a metalens doublet [7].

Figure 5.3 illustrates the off-axis aberration correction resulting from implementing the phase profile described by equation (5.2) in a metalens with a diameter of 20 μm and NA = 0.8 [6]. The figure includes the case of an equivalent metalens with a hyperbolic phase profile for comparison. Figure 5.3 shows field diagrams of focused light by two metalenses, both with $f = 15$ μm and $\lambda = 532$ nm, whose phase profiles are: (a)–(c) hyperbolic, and (d)–(f) optimized (as given by equation (5.2)). In (a) and (d) light is incident at normal incidence (not shown), and it is focused at a focal distance of ~15 μm. In (b) and (e) light is obliquely incident. At normal incidence, the hyperbolic profile produces an aberration-free focal spot in (a), while the optimized profile shows slight on-axis aberration in (d). However, at oblique incidence, when the angle of incidence is not 0°, the focused light by the hyperbolic

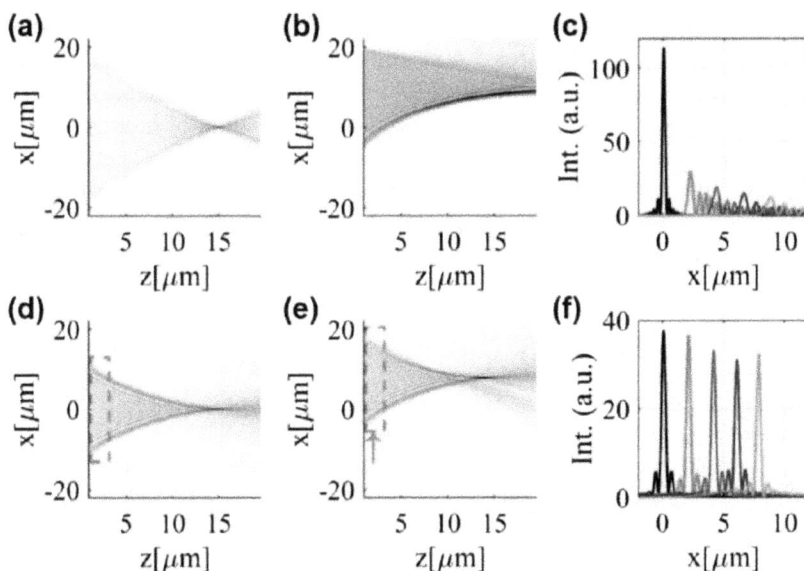

Figure 5.3. Diffraction diagrams showing off-axis aberration correction. The figure shows diagrams of light focusing for two metalenses ($f = 15$ μm, $\lambda = 532$ nm) with phase profiles: (a)–(c) hyperbolic, and (d)–(f) optimized (equation (5.2)). The first and second columns depict the field distributions in the xz-plane for normal (0°) and oblique (30°) incidences, respectively. The metalens is situated on the transverse plane (xy-plane). The third column displays the focal spot intensities for different angles of incidence (depicted in colors). It shows the focused intensity along the x-direction at the focal plane at 0° (black), 7.5° (red), 15° (green), 22.5° (blue), and 30° (orange). The red dashed boxes in (d) and (e) indicate the presence of an effective aperture associated with the optimized phase profile. Reprinted with permissions from [6], copyright 2019 Optical Society of America.

profile is strongly distorted (b), while the optimized profile produces an aberration-reduced focal spot (e). This off-axis aberration is depicted as point spread functions (PSFs) in figures 5.3(c) and (f) for five angles of incidence (indicated in colors). These off-axis (mainly coma) aberrations severely limit the field of view, which is one of the main challenges in designing high-NA metalenses. But indeed, this difficulty of correcting both spherical and off-axis aberrations simultaneously is an issue for all types of lenses when using a single lens. In the examples of figure 5.3 it is illustrated by the hyperbolic phase profile, which corrects spherical aberrations but adds coma, while the optimized phase profile does the opposite.

Using a single metalens, the numerical optimization of the hyperbolic phase profile (equation (5.2)) converges to the so-called spherical phase profile [6]. It is the phase profile equivalent to a spherical lens, given by:

$$\Phi(r) = -k\frac{R_l}{f}(R_l - \sqrt{R_l^2 - r^2}), \tag{5.4}$$

where R_l is the radius of curvature of an equivalent spherical lens. This phase profile produces slightly less off-axis aberration than the numerically optimized phase profile. Figure 5.4 illustrates the off-axis aberration correction resulting from implementing the spherical phase profile described by equation (5.4), this metalens has the same optical parameters ($f = 15$ µm, λ=532 nm) as in figure 5.3, for comparison. Light is incident at normal incidence (not shown), and it is focused at a focal distance of ~15 µm. Figures 5.4(a) and (b) show the field distributions in the xz-plane for normal (0°) and oblique (30°) incidences, respectively. Figure 5.4(c) shows that the focal spot is nearly unchanged with the angle of incidence (indicated in colors), so the metalens is free of coma.

Based on a comprehensive phase profile analysis, an aberration-free metalens design for a wide range of incident angles was reported in [8], where a flat lens design method applies an aperture stop in front of a metalens to achieve aberration-free

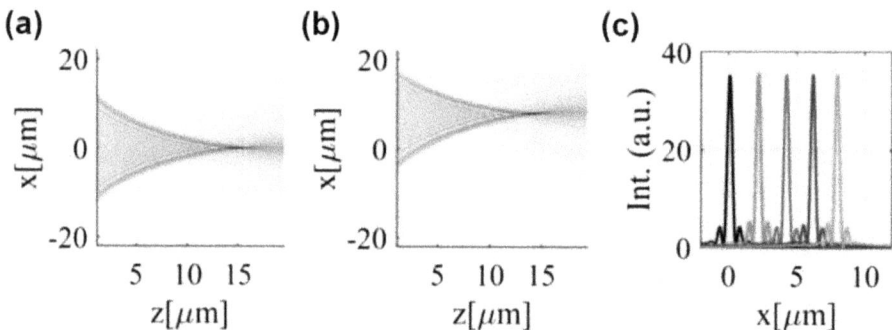

Figure 5.4. Diffraction diagrams showing off-axis aberration correction for metalenses ($f = 15$ µm, $\lambda = 532$ nm) with a spherical phase profile (equation (5.4)). The first and second columns depict the field distributions in the xz-plane for normal (0°) and oblique (30°) incidences, respectively. The third column displays the focal spot intensities for different angles of incidence (depicted in colors), at angles: 0° (black), 7.5° (red), 15° (green), 22.5° (blue), and 30° (orange). Reprinted with permissions from [6], copyright 2019 Optical Society of America.

behavior for both normal and oblique incidence. In this analysis, only parts of the metalens are illuminated with aperture stops to achieve aberration-free focusing. Their analysis also shows that the classical hyperboloidal phase profile is optimal only for normal incidence, and a new sinusoidal phase profile is proposed for better off-axis illumination.

Ultimately, the phase profile that the fabricated metalens imparts to the incident light is the important target. Undersampling the phase profile of metalenses, and fabrication imperfections may contribute to the difference between the desired and the experimental phase profiles. Optimizing the unit cell size of the array of meta-atoms, and developing a more precise fabrication process can increase the match between the simulated and experimental phase profiles.

5.4 Array of meta-atoms

The third step of metalens design (figure 5.1(c)) involves replicating the target phase profile $\Phi(x,y)$ by selecting meta-atoms from a library to impart phase shifts, such that all meta-atoms together match $\Phi(x,y)$ as closely as possible. For example, the circular pillars must have larger diameters towards the metalens center and thinner diameters towards the periphery, and this pattern is repeated periodically because the phase profile wraps back every 2π (see section 2.2.4). Similarly, in geometric-phase metalenses, the nanofins are rotated periodically in zonal discs. This task can be implemented and optimized through an algorithm that repeatedly identifies the required meta-atom for every metasurface coordinate (x,y). However, the size of a metalens may range from millimeters to centimeters, and due to the subwavelength size and spacing of nanostructures, the number of constituent meta-atoms can be very large. Therefore, compression algorithms are required to generate layout files for nanolithographic fabrication. Geometric symmetries and advanced coding languages are utilized to compress the design files of centimetre-scale metalenses from gigabytes to only megabytes [9].

5.4.1 Array geometries

A matrix for positioning meta-atoms must be chosen. Typically, the metasurface is a subwavelength-spaced array of nanoelements in a hexagonal or square grid (figure 5.5), each with adjustable diameter; or with adjustable shape and orientation if they are nano-antennas or nanofins. The adjustable meta-atom parameter is run in a series of simulations that sweep through the possible values (library), in order to select a set of adjustable parameters where the phase shift varies across the whole 0–2π range, while the transmitted intensity is maximized. To match the desired phase at each location, the data of library is used to determine the best nanopillar diameter for creating the phase profile.

The hexagonal and square matrices are periodic arrays of meta-atoms, i.e., their positions are equidistant. These arrays are suitable for low-NA metalenses, or for a high-NA if the metasurface has a wide subwavelength spacing. If unit cells are large or the refractive index is high, meta-atoms do not interact much with their neighbors. This is usual in NIR dielectric metalenses, because the available materials

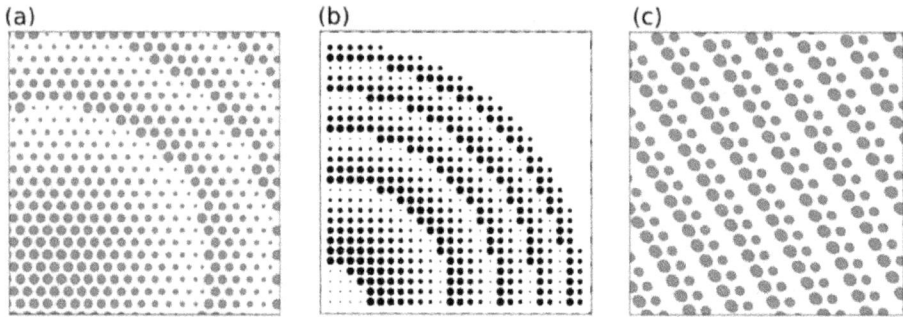

Figure 5.5. A typical arrays of nanopillars: (a) hexagonal array, (b) square array, and (c) radial array. (a) and (c) Reprinted with permission from [10], copyright 2016 Optical Society of America. (b) Adapted by permission from Springer Nature Service Centre GmbH [Nature] [Nature Reviews Materials] [1], copyright (2020).

have a high refractive index (amorphous Si, $n \sim 3.6$), confining NIR light well, leading to less interactions between neighboring pillars [10].

Large metalenses (mm- to cm-scale diameters) may be quite difficult to simulate and optimize, due to their large area. This difficulty is especially pronounced at the periphery of high-NA visible-light metalenses, where hexagonal and square arrays are locally inaccurate due to strong phase gradients, interactions between neighboring pillars, and large oblique angles of light. In addition, hexagonal and rectangular grids lead to arrays with very little symmetry or order away from the center (metalens periphery), where pillars with equal diameter become non-periodic among them. This issue can be addressed by employing a different grid geometry. There are methods for designing a large-area and high-NA metalens that use a meta-atom array with radial symmetry (figure 5.5(c)). This radial array allows a high optical efficiency via the optimization of the adjustable parameter and position of each individual meta-atom, with a computational cost that is nearly independent of the metalens size [10]. Because the basic metalens phase profile has circular symmetry, the radial symmetric grid (figure 5.5(c)) is very suitable. For example, a millimeter-scale metalens might have 100,000-fold rotational symmetry, while the hexagonal grid only has D6 symmetry, and the square grid only has D4 symmetry. In addition, away from the metalens center, the phase profile changes rapidly (approximately linearly), approaching a periodic sawtooth phase. In the hexagonal and square arrays, the pillars are placed on a fixed-period lattice, but the sawtooth phase profile has a non-periodic phase sampling. The radial grid allows the pillars to be at arbitrary radial and azimuthal locations within 2π phase periods. Thus, the radial array matches the phase sampling required by the sawtooth phase profile.

In summary, meta-atoms in a radial grid locally approximate the hyperboloidal phase profile as a linear phase profile in metalenses with high-NA. The optimal grid is a combination of both the hexagonal and the radial arrays, with the hexagonal grid in the center of the metalens, and the radial grid elsewhere. As one approaches the center of the lens, the hexagonal array becomes progressively better than the radial because there is no sawtooth phase. Conversely, the hexagonal matrix worsens towards the periphery. The recommended switch between the two grid geometries is at ~f/4 distance (where light bends by 15°) from the metalens center, where f is the focal length. In optimized designs of metalenses with NA near 1.0, the hexagonal grid may be only 1% of the metalens area, and the remaining 99% be a radial grid [10].

5.4.2 Numerical aperture limit

The NA of a metalens is limited by the size of the unit cell of their meta-atoms [11]. Note that the phase profile $\Phi(r)$ is discretely reproduced due to the finite size p of unit cells, this adds distortion to the focused wavefront, and its effect increases with the diameter of the metalens, which in turn limits the maximum NA. This can be evaluated by the Nyquist–Shannon sampling theorem in the spatial domain by considering the periodicity of the unit cell p as the sampling size [11, 12]. According to the Nyquist–Shannon theorem, the spherical aberration could be prevented when NA $\leqslant \lambda/(2p)$. In other words, to prevent spherical aberration, the metalens unit cell must satisfy the following condition [11]:

$$p \leqslant \frac{\lambda}{2\mathrm{NA}}. \tag{5.5}$$

For example, a metalens designed at $\lambda = 532$ nm and NA $= 0.8$ must have a unit cell periodicity $p \leqslant 332.5$ nm. The larger the NA, the smaller the required unit cell, but the higher the efficiency of the metalens due to increased sampling. Conversely, the smaller the unit cell, the larger the achievable NA, and higher the metalens efficiency. In other words, small periodicities (high sampling rates) are usually set for large NA metalenses [12]. However, for a given set of meta-atom parameters (material, height, etc), the phase shift range (2π) may decrease as p decreases. To maintain the optical performance at the operation wavelength, one is required to modify the other parameters such as the meta-atom height.

Example A metalens has a diameter of 20 µm and a focal length of 10 µm in air. When using a wavelength of 600 nm what will be its (a) acceptance angle, (b) numerical aperture, and (c) the maximum unit cell size?
Solution

(a) Since $\sin \theta_{max} = D/2f$,

$$\theta_{max} = \sin^{-1}(20/2 \times 10) = 45°$$

Hence

$$2\theta_{max} = 90°$$

(b) From NA $= n_i \sin\theta_{max}$

$$\mathrm{NA} = (1)\sin(45°) = 0.707$$

(c) From equation (4.1)

$$p \leqslant 600 \text{ nm}/(2*0.707)$$

$$p_{\text{max}} = 424 \text{ nm}$$

5.4.3 Unit cell

An important feature for metalenses is surface discretization, which is determined by the characteristics of the unit cell: size (spacing between adjacent meta-atoms), geometry, and its composition (meta-atom or meta-molecule). Figure 5.6 shows four unit cells: square shape, hexagonal shape, trapezoidal shape, and a two-atom unit cell. The unit cell size, denoted by p, must satisfy the sampling (Nyquist) criterion to ensure diffraction-limited focusing as described in equation (5.5). As explained in section 5.4.2, satisfying this criterion becomes more challenging for high-NA metalenses. However, this is not the only criterion for the unit cell size, as it must also be subwavelength. In other words, its size must be $p < \lambda$ to avoid unwanted

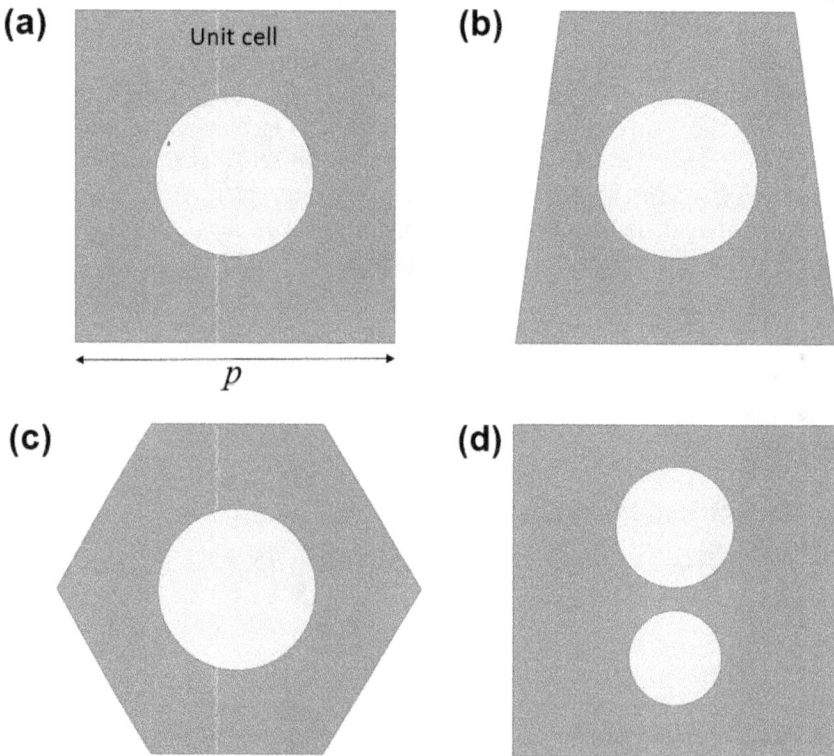

Figure 5.6. Different unit cells with circular meta-atoms (yellow color): (a) square unit cell with size p, (b) trapezoidal shaped unit cell, (c) hexagonal unit cell, and (d) a supercell with two meta-atoms.

diffraction orders. On the other hand, the unit cell size cannot be too small. For nanopillars, the unit cell size must be large enough to support all pillar sizes required to achieve full phase coverage. Also, because the nanopillar height is fixed, as p becomes smaller higher aspect ratios are required, making their fabrication more difficult. Moreover, smaller unit cells bring adjacent meta-atoms closer together, reducing light confinement through meta-atoms and increasing field coupling between adjacent meta-atoms.

Unit cells can be more complicated structures, consisting of meta-molecules instead of meta-atoms (figure 5.6(d)). These unit cells are supercells consisting of multiple different features, designed to achieve higher optical performance. One example is a two-atom meta-molecule [13], which consists of an asymmetric dimer nanoantenna made of two nanodisks of different diameters, separated by a short distance. These asymmetric dimer nano-antennas can generate asymmetric radiation patterns to efficiently bend incident light at large angles (up to 80°). These supercells can address some metalens issues, but at the cost of a larger unit cell size, which still must satisfy equation (5.5).

5.5 Chromatic aberrations

The quest for broadband achromatic metalenses, capable of maintaining a single focal length across a wide spectrum of wavelengths, represents a significant challenge due to the difficulty in engineering an invariant phase profile at distinct wavelengths.

Historically, the challenge of chromatic aberrations has proven to be a significant obstacle. It took nearly two centuries to address chromatic aberrations in refractive lenses, beginning with Sir Isaac Newton's reflective telescope in the early 1700s and culminating in the development of the achromatic optical system by Zeiss–Abbe–Schott in the late 1800s [1]. Therefore, it is reasonable to anticipate that solving chromatic aberrations in metalenses will also pose a significant challenge, requiring both perseverance and innovation.

The chromatic aberrations in refractive lenses are controlled by a special combination of glass dispersions in lens doublets [5]. Chromatic aberration L_c of a single thin refractive lens is directly proportional to the lens's focal length f and inversely proportional to the Abbe number V of the glass, given by $L_c \approx -f/V$. A lower Abbe number indicates higher dispersion [5]. It is possible that in metalenses, chromatic aberrations follow a similar relationship with focal length and wavelength-dependent optical effects. In metalenses, these optical effects may include dispersion, scattering, and diffraction. In refractive lenses, material dispersion is recognized as the primary factor influencing chromatic aberration, with surface curvatures having minimal impact [5]. However, in metalenses, both wavelength-dependent effects and surface profile play a role. Dispersion and scattering are due to meta-atoms, and the spatial profile of meta-atoms contributes to diffraction. This wavelength-dependent diffraction arises from the 'concentric diffraction grating' formed by the wrapping of Fresnel zones in the array of meta-atoms, as shown in

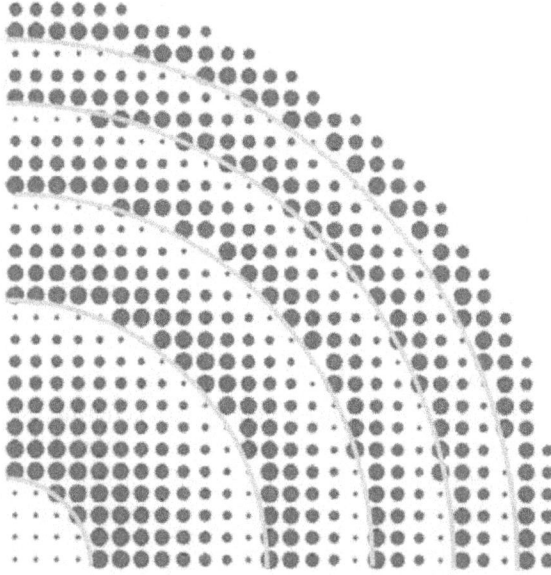

Figure 5.7. Fresnel zones in a metalens, its separation indicated by yellow lines. The pattern of meta-atoms is repeated periodically because the phase profile wraps back every 2π. Adapted by permission from Springer Nature Service Centre GmbH [Nature] [Nature Reviews Materials] [1], copyright (2020).

figure 5.7 [14]. Therefore, an achromatic metalens should minimize the different sources of chromatic aberration or use these sources to cancel each other.

One approach to generalize the different sources of chromatic aberration in metalenses is by expanding the phase profile in a Taylor series, around a design frequency ω_d [15], which is given by

$$\Phi(r, \omega) = \Phi(r, \omega_d) + \frac{\partial \Phi(r, \omega)}{\partial \omega}\bigg|_{\omega=\omega_d} (\omega - \omega_d) + \frac{\partial^2 \Phi(r, \omega)}{2\partial \omega^2}\bigg|_{\omega=\omega_d} (\omega - \omega_d)^2 + \dots , \quad (5.6)$$

where ω_d is the design frequency, $\Phi(r,\omega_d)$ is the target phase profile, and $\Phi(r,\omega)$ is the frequency-dependent phase profile. In general, the metalens exhibits different phase profiles depending on the frequency of incident light. The typical desired phase profile is the parabolic, which can be written as

$$\Phi(r, \omega) = -\frac{\omega}{c}(\sqrt{r^2 + f^2} - f), \quad (5.7)$$

where ω_d is the design frequency, c is the light speed, and r and f are the radial coordinate and focal length, respectively. It's important to note that, in general, equation (5.7) is only valid for a single frequency, as the shape of the phase profile may change with frequency, making it perfectly suitable to evaluate $\Phi(r,\omega_d)$ in equation (5.6). If the profile considerably changes with frequency, may not to be a focal point at all. If the change with frequency is small, one could consider using the same equation (5.7), but with a slight change in the focal length as a function of frequency, i.e. $f(\omega)$. Note that equation (5.7) is essentially achromatic, because the light control is due to the phase gradient term in the refraction law equation (2.1), which is not frequency dependent:

$$\frac{c}{\omega_d}\frac{\partial\Phi(r,\omega_d)}{\partial r} = -r\left(\sqrt{r^2+f_d^2}\right)^{-1},\qquad(5.8)$$

where f_d is the design focal length. Therefore, a hyperbolic metalens is achromatic if its equation (5.7) is valid for different values of ω_d.

The direct chromaticity correction would involve canceling or minimizing the derivative terms in equation (5.6), with the ideal case being $\Phi(r,\omega) = \Phi(r,\omega_d)$. However, correcting metalens chromatic aberration is more complex than that. Reducing chromatic aberration involves employing various dispersion engineering techniques on the meta-atoms [1, 14, 15].

The first term of equation (5.6) is the design phase profile $\Phi(r,\omega_d)$, which produces a spherical transmitted wavefront when the target phase is equation (5.7). To achieve achromatic focusing with a given bandwidth $\Delta\omega$ around the design frequency ω_d (usually chosen at the center of $\Delta\omega$), all meta-atoms must address not only $\Phi(r,\omega_d)$, but also the higher-order derivative terms. The first-order and second-order derivative terms, $\partial\Phi/\partial\omega$ and $\partial^2\Phi/\partial\omega^2$, are the group delay and group delay-dispersion profile, respectively. These optical parameters are directly related to group velocity $\partial k/\partial\omega$, and group velocity dispersion $\partial^2 k/\partial\omega^2$ [16]. The group delay is the transit time required for traveling a distance at a given group velocity. Since the group delay may itself be frequency-dependent, different frequency components of the light pulse may undergo different group delays, causing the pulse to spread in time. A measure of this spread is the group delay dispersion.

Therefore, these derivative terms strongly determine the chromatic focal-length shift of the metalens. It is desirable for these delays to be constant across all frequencies; otherwise, there is temporal smearing of the focused light. Of course, equation (5.6) has more derivative terms, and the more terms one considers, the better the correction of chromatic aberration. There are several procedures to deal with equation (5.6). One approach designs meta-atoms or meta-molecules in order to combine the variation with (r,ω) to deal with the phase, group delay, and group-delay dispersion simultaneously [15]. In such an approach, the incident light is considered as individual wave packets, where meta-atoms (or meta-molecules) are engineered to tune the time delay differently at the center of the metalens than on the sides, ensuring that all wave packets arrive at the focus at the same time, regardless of their location upon traversing the metalens (figure 5.8). This requires a time delay (group-delay) independent of frequency:

$$\frac{\partial\Phi(r,\omega)}{\partial\omega} = -\frac{1}{c}(\sqrt{r^2+f^2}-f),\qquad(5.9)$$

where f is ideally constant. This ideal condition makes the second-order and other high-order derivatives zero for all coordinates, which leads to equation (5.7), which is essentially achromatic. This ideally ensures that all transmitted light of different frequencies constructively interferes at the focal point. For visible light, group

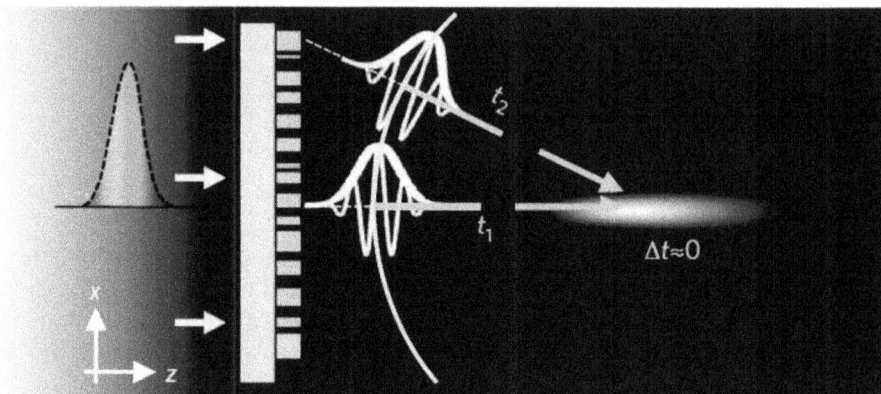

Figure 5.8. To achieve achromatic focusing, the wave packets should reach the metalens focus simultaneously while preserving their temporal profiles. An achromatic metalens can be designed to offer spatially dependent group delays, ensuring that wave packets from any position on the metalens arrive synchronously at the focus. The yellow curve represents the spherical wavefront. Reprinted by permission from Springer Nature Customer Service Centre GmbH: [Nature] [Nature Nanotechnology] [15], copyright (2018).

delay and group delay dispersion are in units of femtoseconds (fs) and squared femtoseconds (fs^2). Chromatic effects can be corrected by considering the first three terms in equation (5.6), by artificially adjusting their values satisfying the following conditions: (a) tuning the phase $\Phi(r,\omega_d)$ to generate a spherical wavefront; (b) using group delay to compensate different wave packet arrival times at the focus; and (c) using group delay dispersion to make identical outgoing wave packets.

However, to take into account the chromatic dispersion of metalens, and to analyze achromatic focusing for a bandwidth, the focal shift must be considered in equation (5.7). A simple model of focal length shift is given by [15]:

$$f(\omega) = f_d \left(\frac{\omega}{\omega_d} \right)^n,$$

(5.10)

where ω_d is the design frequency and f_d is the design focal length. The exponent n is a real number that is related to the degree of dispersion, and then metalens dispersion can be analyzed by substituting different values for n. A metalens with $n = 0$ can be considered achromatic, with no focal shift with frequency. A metalens with $n = 1$ can be considered conventional diffractive, with a focal shift similar to a Fresnel lens. A metalens with $n = -1$ has a focal shift proportional to wavelength, similar to a refractive lens. Equation (5.10) indicates that when n has negative values, longer wavelengths are concentrated farther away from the metalens. Conversely, positive values of n imply that longer wavelengths are concentrated closer to the metalens. As the absolute value of n increases, the stronger the dispersion, and the separation between the focal spots of two wavelengths increases. Figure 5.9 plots the focal shifts, and required group delays and group delay dispersions for metalenses with 20 μm diameter, design focal length $f_d = 50$ μm, and design wavelength $\lambda_d = 530$ nm.

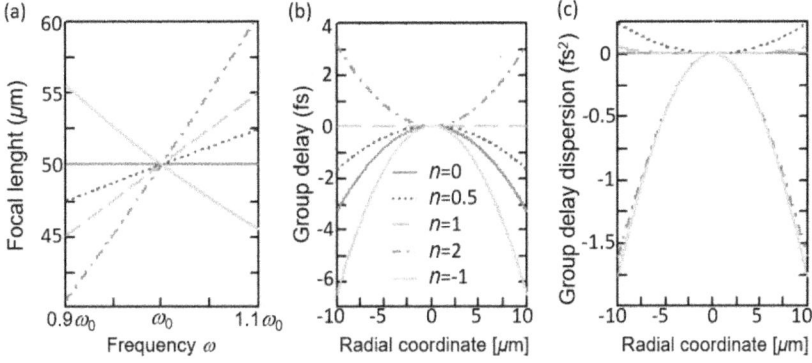

Figure 5.9. Dispersion analysis of metalenses. (a) Focal length shift with frequency, where the focal length is modeled by equation (5.10) through a bandwidth of $\Delta\omega = 0.2\omega_o$. (b) Required group delay as a function of metalens radial coordinate. The metalens can be designed in function of n as: achromatic ($n = 0$), chromatic with focal length inversely proportional to wavelength ($n = 1$), or chromatic with focal length proportional to wavelength ($n = -1$). Values of $n = 0.5$ and 2 are also shown. These metalenses have a diameter of 20 μm, a focal length of 50 μm at a design wavelength $\lambda_d = 530$ nm. (c) Required group delay dispersion for the same metalenses.

Metalenses with n values of 2 and -1 require higher group delay dispersion to precisely control the focal length shift. Conversely, for $n = 1$, the required group delay and group delay dispersions are comparatively small.

To design an achromatic metalens, the phase spectrum of each meta-atom must be computed in a library at the design wavelength λ_d (or design frequency ω_d) within a bandwidth $\Delta\lambda$ (or $\Delta\omega$) to obtain the group delay. A fit of the phase spectrum of selected meta-atoms from the library must meet the requirement for realizing achromatic metalenses within the design bandwidth. FDTD simulations must be conducted to demonstrate that the metalens remains achromatic up to the design bandwidth. To achieve the required group delay for each point of the achromatic metalens, the appropriate meta-atom is positioned with a suitable phase shift, ensuring the necessary phase profile at the design wavelength λ_d. The value n is determined for the metalens, and the phase spectrum is approximated with a polynomial of n order to determine the group delay and group delay dispersion.

There are several design principles to achieve broadband achromatic metalenses. Another method incorporates an additional phase shift to the phase profile of the metalens, given by [17–19]:

$$\Phi(r, \lambda) = -\frac{2\pi}{\lambda}(\sqrt{r^2 + f^2} - f) + C(\lambda), \qquad (5.11)$$

This strategy for achieving a broadband involves introducing a wavelength-dependent phase shift $C(\lambda)$ to compensate for the spectral phase variations. For example, broadband achromatic metalenses in reflection mode (1200–1680 nm), and in transmission mode (400–660 nm) were developed in [17, 18]. To conserve a linear relation with $1/\lambda$, a phase shift inversely proportional to the wavelength, given by $C(\lambda) = \alpha/\lambda + \beta$, was employed, where $\alpha = \chi \lambda_{max} \lambda_{min}/(\lambda_{max} - \lambda_{min})$ and $\beta = -\chi \lambda_{min}/(\lambda_{max} - \lambda_{min})$, with λ_{min} and λ_{max} representing the shorter and longer wavelengths, and χ indicating the largest additional phase shift C_{max} between λ_{min} and λ_{max} at the central position of the metalens. The phase shift between the maximum and minimum wavelengths is compensated by the spectral responses in the meta-atoms. The spectral term can also be expressed as a function of frequency $C(\omega)$.

5.6 Focusing efficiency

Simulated or measured focal spots and their corresponding horizontal cuts indicate the focusing quality of the designed or fabricated metalens. Focusing efficiency quantifies such focusing quality and is calculated as the ratio of the optical power of the focused beam to that of the incident beam. The optical power of the incident beam is calculated as the optical power passing through the optical aperture, usually circular, typically with the same diameter as the metalens. The optical power of the focused beam is calculated as the area of the central maximum of the intensity profile in the focal plane. In other words, the focusing efficiency is the ratio of the optical power in the focal spot area (central area of diameter approximately 2× FWHM) to the incident optical power. In most cases, the focusing efficiency depends on the numerical aperture (NA) and the design wavelength of the metalens.

Another figure of merit is the Strehl ratio, which measures the quality of focusing and image formation, with values ranging between 0 and 1. A perfectly unaberrated metalens has a Strehl ratio of 1, but common practice considers a lens to be 'diffraction-limited' when its Strehl ratio is greater than 0.8. To calculate this ratio, the intensity profile at the focal plane is normalized to that of the ideal Airy function with the same area under the curve. The Strehl ratio is the ratio of the maximum intensity of the focal spot (of the evaluated metalens) to the ideal maximum intensity from a theoretically diffraction-limited equivalent lens. For this purpose, an ideal Airy function is overlaid onto each metalens' intensity profile at the focal plane.

The simplest way to assess the focusing quality is through the full width at half-maximum (FWHM) of the focused intensity profile. For example, if the focal plane is the xy-plane, the FWHM is the Δx difference between the two values ($\pm$) of x at which the intensity is equal to half of its maximum value. In other words, it is the width of the focused intensity curve measured between those points on the x-axis (or y-axis) which are half the maximum intensity. A perfect or diffraction-limited lens has a FWHM approximately equal to $0.5\lambda/NA$.

The modulation transfer function (MTF) of a metalens is an advanced metric for assessing imaging quality. The MTF determines how much spatial contrast of the original object is maintained in the formed image, characterizing the spatial frequency content in the image with respect to the object.

5.7 Problems

5.1 **Chromatic Aberration in Metalenses**. Analyze the chromatic aberration for a metalens designed to focus light at wavelength λ_0. Determine the focal shift for wavelengths λ_1 and λ_2, where $\lambda_1 < \lambda_0 < \lambda_2$. Assume the phase profile of the metalens is optimized for λ_0.

5.2 **Polarization-Dependent Focusing**. Design a metalens that focuses differently for different polarization states of incident light. Derive the phase profile required for the metalens to have a focal length f_E for TE polarization and a different focal length f_M for TM polarization.

5.3 **Metalens with Spherical Aberration**. Consider a metalens with a spherical phase profile that introduces spherical aberration. Derive the expression for the wavefront error and the resulting intensity distribution at the focal plane. Discuss how the aberration affects the resolution and image quality.

5.4 **Metalens Design**. Design a metalens with a high numerical aperture (NA) for focusing light at a wavelength λ. The desired NA is 0.9. Select an array geometry for the meta-atoms that satisfies the phase profile while maintaining structural suitability and manufacturability.

Simulate the focusing behavior of the metalens using electromagnetic simulation software. Optimize the design to maximize focusing efficiency and minimize aberrations.

5.5 **Chromatic Aberration Minimization**. Design a metalens that minimizes chromatic aberration for a wavelength range from $\lambda_1 = 500$ nm to $\lambda_2 = 700$ nm. Derive the phase profiles needed to focus different wavelengths (λ_1 and λ_2) to the same focal point. Propose a design for meta-atoms that can achieve the required phase shifts for multiple wavelengths.

Describe an experimental method to measure the performance of the metalens in terms of chromatic aberration reduction.

5.6 **Focusing Efficiency Optimization**. Analyze and optimize the focusing efficiency of a metalens designed to focus light at a wavelength λ with a given numerical aperture (NA). Propose an expression for focusing efficiency considering the transmission, phase coverage, and losses (scattering, diffraction, absorption). Design an array with optimized spacing and arrangement to ensure uniform phase coverage to maximize transmission and minimize scattering, using high-index dielectrics or different types of meta-atoms.

Bibliography

[1] Chen W T, Zhu A Y and Capasso F 2020 Flat optics with dispersion-engineered metasurfaces *Nat. Rev. Mater.* **5** 604–20

[2] Moon S-W, Lee C, Yang Y, Kim J, Badloe T, Jung C, Yoon G and Rho J 2022 Tutorial on metalenses for advanced flat optics: design, fabrication, and critical considerations *J. Appl. Phys.* **131** 091101

[3] Oh B, Kim K, Lee D and Rho J 2023 Engineering metalenses for planar optics and acoustics *Mater. Today Phys.* **39** 101273

[4] Hecht E 2002 *Optics* (Reading, MA: Addison-Wesley)

[5] Kingslake R and Johnson R B 2010 *Lens Design Fundamentals* 2nd edn (New York: Academic/SPIE)

[6] Liang H, Martins A, Borges B-H V, Zhou J, Martins E R, Li J and Krauss T F 2019 High performance metalenses: numerical aperture, aberrations, chromaticity, and trade-offs *Optica* **6** 1461–70

[7] Groever B, Chen W T and Capasso F 2017 Meta-lens doublet in the visible region *Nano Lett.* **17** 4902–7

[8] Kalvach A and Szabó Z 2016 Aberration-free flat lens design for a wide range of incident angles *J. Opt. Soc. Am.* B **33** A66–71

[9] She A, Zhang S, Shian S, Clarke D R and Capasso F 2018 Large area metalenses: design, characterization, and mass manufacturing *Opt. Express* **26** 1573–85

[10] Byrnes S J, Lenef A, Aieta F and Capasso F 2016 Designing large, high-efficiency, high-numerical-aperture, transmissive meta-lenses for visible light *Opt. Express* **24** 5110–24

[11] Chen W T, Zhu A Y, Khorasaninejad M, Shi Z, Sanjeev V and Capasso F 2017 Immersion meta-lenses at visible wavelengths for nanoscale imaging *Nano Lett.* **17** 3188–94

[12] Pan M, Fu Y, Zheng M, Chen H, Zang Y, Duan H, Li Q, Qiu M and Hu Y 2022 Dielectric metalens for miniaturized imaging systems: progress and challenges *Light Sci. Appl.* **11** 195

[13] Paniagua-Domínguez R *et al* 2018 A metalens with a near-unity numerical aperture *Nano Lett.* **18** 2124–32

[14] Arbabi E, Arbabi A, Kamali S M, Horie Y and Faraon A 2016 Multiwavelength polarization-insensitive lenses based on dielectric metasurfaces with meta-molecules *Optica* **3** 628–33

[15] Chen W T, Zhu A Y, Sanjeev V, Khorasaninejad M, Shi Z, Lee E and Capasso F 2018 A broadband achromatic metalens for focusing and imaging in the visible *Nat. Nanotechnol.* **13** 220–6

[16] Saleh B E A and Teich M C 1991 *Fundamentals of Photonics* (New York: Wiley)

[17] Wang S *et al* 2017 Broadband achromatic optical metasurface devices *Nat. Commun.* **8** 187

[18] Wang S *et al* 2018 A broadband achromatic metalens in the visible *Nat. Nanotechnol.* **13** 227

[19] Aieta F, Kats M A, Genevet P and Capasso F 2015 Multiwavelength achromatic metasurfaces by dispersive phase compensation *Science* **347** 1342–5